Connected

Walking with Nature

Connected

Walking with Nature

Jonathan Davidson

Troubador Publishing Ltd
Unit E2 Airfield Business Park
Harrison Road, Market Harborough
Leicestershire LE16 7UL
Tel: 0116 279 2299
Email: books@troubador.co.uk
Web: www.troubador.co.uk

ISBN 978 1 80514 166 2

British Library Cataloguing in Publication Data.
A catalogue record for this book is available from the British Library.

Printed and bound in Great Britain by 4edge Limited
Typeset in 12pt Minion Pro by Troubador Publishing Ltd, Leicester, UK

Matador is an imprint of Troubador Publishing Ltd

For Hallie, Reuben, Noah, Immy, Cassius,
Louis, Luca, Matilda and Fede

Contents

Contents

Introduction

These stories reflect the great passion I have for the natural world and will hopefully inspire others to explore the gift of the natural world that surrounds us. I call the collection *Connected* as this is how I truly feel when I am out walking, when everything links together, and I become one with the natural world around me. There are the elements, sounds, smells and wide variety of sights you get when in the countryside. To recognise your place without the need to dominate is both humbling and absorbing. I feel blessed that at an early age my interest in the natural world blossomed as, without it, I would feel left wanting.

I also choose the word *connected* as there is a train of thought within the world of environmentalists that fewer and fewer people are familiarising themselves with the natural world, and I share this view. If our worst fears are realised, then this disconnect will mean the natural world becoming marginalised, resulting in its increasing deterioration. This process will become more acute and more urbanised as the human population grows. Somehow, we must find the key to helping people understand not only the benefits of nature but also the immense joy it can bring.

One of my earliest memories of the natural world started with humble beginnings, and for that I thank my mother, for hanging out a red basket of peanuts which in turn attracted a host of birds, predominantly blue tits. I was nine at the time and would sit at the kitchen table transfixed by these active and charismatic creatures. The following spring, I noticed a pair of blue tits going to and from a crack in a wall with nesting material and later still, listening to the young calling from within. I was hooked.

The world opened to me, and I wanted to know more. Initially my interest stayed with birds as I wanted to be able to identify them visually and by song. I bought bird books, reading them cover to cover, and was given a pair of binoculars as a birthday present. At the time I lived in Kingsbridge, South Devon so would walk the estuary testing my new-found skills. My other great explorative region was my grandparents' fifty-acre farm in Whimple, east of Exeter, a small farm by today's standards and managed sensitively and not in competition with nature. Beautiful hay meadows bordered by hedgerows that, in the spring, were covered in blooms from a diverse number of plants and, in the autumn, a bountiful harvest of berries that carried the wildlife through the winter months. An abundance of tall mature trees standing across the farm and the watery goyle attracting all sorts of wildlife. Growing up, I had the freedom to explore the land looking for any number of farmland wildlife.

As I got into my teens, we moved to Thurlestone Rock where I had a bedroom overlooking low-lying land that flooded during the winter. This turned into my private reserve filled with wigeon, teal, snipe and several other waterfowl and waders. It was a wonderful sight and just full of activity. It was on this marsh that I first encountered one of my other great wildlife passions: the brown hare. In the spring they would take to the slightly higher ground to congregate and court which occasionally turned into a bout of boxing.

With my early mastering of birdwatching and the introduction of the brown hare, it started to dawn on me that everything was interdependent (what we now call biodiversity), and this in turn led me to want to know more about habitat, animal specialities and who survived where and by what means. The web of life was beginning to take shape.

The natural world has always been immensely important to me but equally, so has my art and photography. From my earliest memories I can recall the pleasure of creating paintings and photographing landscapes and animals. There has never been a line defining the three interests.

With the acquisition of my bird books, I wanted to understand the birds' profiles, colours and characteristics, and so I would spend hours copying and painting the birds from my books. As well as the pure pleasure of painting and creating, this copying process helped me get closer to understanding the beauty of the birds and how to identify them. As I got older, so my repertoire extended to landscape painting, in many cases including wildlife.

My enjoyment in photography started at a very early

age. My father loved his photography, developing and printing his own images. As a small boy I would spend time with my father in the darkroom where I was put in charge of the red and white lights, ensuring I turned them on at the critical moment in the printing process. I learned a lot about tonality and composition during those informative times in the darkroom. Over subsequent years I have used my artistic inspiration to develop my wildlife and landscape photography. The joy of taking a great landscape or wildlife photograph always feels like a tribute to the location and its nature.

From this crucial step to the present day, I have been on a great journey of discovery. When there are no days of enlightenment there is always the familiar, which still makes me smile and something I never tire of.

Early Years

Brown Hare

Spring was arriving, and with the draining of the marsh and increasing temperatures, the land had quickly thrown up a sward of rich grasses and wild herbs, and it was on this lush expanse that I saw my first brown hares. The marsh is traversed by horizontal drainage ditches and backs onto a beach, protected by a steep dune. Stitched to this tamed, yet rebellious, landscape is a grey sky of increasing brightness but flat in its bearing.

These hares revealed themselves to me when, as a boy, I awoke opening my morning curtains onto my very own wildlife sanctuary. Having lost the company of the overwintering wildfowl and the snipe, I would look across the marsh for signs of spring and anything of interest before I set off on my one-mile walk to the bus stop. And there they were, an initial three animals joined by three more that set about chasing each other, which eventually led to a bout of boxing, and my heart nearly bounced as high as those wonderful athletes as they took to the air.

There is much written about the brown hare, with each author trying to understand and explain what is so special about the animal. In a lot of writing on any animal, the term 'captured' is often used, however with the brown hare, this is not possible as the very spirit of the creature prevents you from using the word, and it is this quality that elevates the brown hare.

Thinking of those early sightings, it is now clear to me that as well as the spirit of the animal, it is also its energy and enthusiasm for life that captured me. When those brown hares on the marsh dashed and weaved, it was as if they were possessing the very wind, tying it down with a magic thread. The hares appeared to have company that they tried to avoid as they swerved something unseen, and it is this mystery of their other, unseen world that leads you to believe that they are inhabiting more than one dimension, skipping between two or more.

Years on, and I am sitting in a lay-by on the A36 in the comfort of my car, which doubles up as a useful hide, and watching for the third season a familiar community of hares. Lorries and cars roar by and are oblivious to the motives, tensions and livelihoods of the gathering. When a tractor comes up the hill, the flighty creatures recognise the differing decibels and react to a familiar intruder driving them from their hiding and over the horizon.

For a lot of the time the hares are sedentary, resting on the broad, sweeping slopes of the endless field in either loose groupings or on their own. I sit in my car, buffeted by the noisy and speeding vehicles on one side and a silent lack of activity on the other, bridging the world I am born to and looking in on one I admire. As the light level drops and the traffic leaves my conscience, I am drawn to the world of the brown hare: a mixture of slumbered awakening, testing of stiffened limbs through stretch exercises and the greeting of fellow. Slowly, the occasional animal will move to a fresh location and try a blade or two of the sprouting winter barley, young fresh shoots, ideal for breakfast. These early movements are ungainly and belie the true physiology of the athlete, born to sprint with its extra-large heart and oxygen-rich blood: this is the warm-up. Once these sprinters of the plain truly awake to their expansive landscape, so they begin to inhabit the never-ending and sweeping blue sky, rising to it in balletic prose.

There is a slow build-up to any bout of boxing. Often the jack will stay close to the jill, never more than a metre from her side. The jack needs to make his presence felt, a constant reminder to the jill and other prospective jacks. This overbearing lover becomes an irritant to the jill, and so she moves away with more determination and enters into a trot up the field. However, he is not to be so easily shaken off and follows in close proximity. The trot turns to a canter, and this is the jack's cue to assert himself, so he closes in and, whilst running at considerable speed, outstretches one paw to clip her back leg. Affronted, the jill spins to confront the arrogance and to give him the brush-off. The jack, unaware of the jill's sudden juddering halt, almost collides with her, and this is her opportunity to cuff him, which she does with force and relish. Now they spin, each standing on all fours a metre apart, the jack distinguished by his erect ears, lending height and buoyed dominance whilst the jill takes a low, menacing approach with ears flattened back: an even match. Using his powerful back legs, the jack rises and moves forward in one fluid motion, and without hesitation, the jill rises to meet him, and so they come together. Thankfully, the action appears in slow motion and appears to be happening over the great arc of time; in reality, it is less than a minute.

The jill has put her suitor to the test, and he must resit the exam which, undeterred, he will happily do. This early action has stirred the community, and there are hares running from one to the other like gossips crossing the street to share stories and embellish on what they have just witnessed. The protagonists disappear over the crest of the hill and are followed by the crowd. You can only imagine the jack will have more than the jill to contend with during the next encounter.

There is a romance to hares that hints they are more than mere mammals looking to extend their gene pool. As they inherit this prehistoric land with its open vistas and barrows of the ancients, an unbroken link to the past, so their wonderment will pass to generations not yet born.

Illuminations

The party is over and it's time to make my way home. It is a July evening, and I have joined school friends at one of their homes for a teenage gathering, conveniently facilitated by the absence of parents and, as it is approaching midnight, some of us start to drift away. The party is in the village of Galmpton, and I live in the nearby hamlet of Thurlestone Rock which is approximately one mile away; both are in a quiet corner of Devon which has yet to experience the influx of visitors eager to share the beauty of the South Hams.

As I leave the hubbub of the party, I enter a world of complete darkness and quietude overseen by a sky that is fractured like a poorly glazed pot with the occasional star forcing an appearance. It is quite a shock to the senses to walk from light to such extreme dark when it is normal to seek solace in light from the pitch black of night. My guides for my journey home are the tall, deep Devon hedges which envelope me and keep me safe as I travel down their weaving path. These tall ramparts are dense with new growth where the primrose has given way to the red campion, flowering blackberries and tall spires of foxglove. The oak, ash and elderberry strike out, not expecting to be flailed into submission during the autumn tidy-up which comes later in the year. In the distance I can hear the whispered hush of the sea as it draws breath and brings me closer to home.

I am making good time and now totally relaxed in my surroundings with many little sounds and rustlings coming from within and beyond my tall escorts. Passing a dilapidated gate, I am confronted by the strong smell of a dog fox that is keen to let me and, more importantly, his kin know of his presence, and although the wily creature and I both stride out with purpose, I sense he has more in common with his surroundings. Having travelled the top road, I prepare to drop down onto the middle road where the hedges are even deeper and taller. When I reach the three-way junction, I would normally stop to take in the view of Thurlestone bay with the dominant rock standing tall and proud and adorned with cormorants drying their wings, but the inkiness keeps me from the pleasure.

Halfway down the middle road a sheep coughs, clearing her throat, and beyond my guiding fortifications her sisters can be heard in their rummaging. As I turn one more bend with all my senses dilated, I come across a battery of lights that resemble blue-green emeralds, and I am pulled up short. I stand back, unsure of what I am looking at, when it dawns upon me that these dozens of little illuminations are glow-worms shining out and flirting with unseen suitors.

On closer inspection, the small creatures appear towards the end of the branches and quite still but remain as part of a close community. The light given off is soft, yet sufficient to see the detail of the hedge around them, and it is easy to see why, in the past, villagers and locals have used them for way finding. They are not worms but beetles about 25mm long, and it is the wingless female that glows in the dark, trying to attract her flighty mate. Once mated, the female extinguishes her light, lays her eggs and turns to death. It is heartening to know that decades later, these unassuming little creatures earned a posthumous award as animal war heroes for assisting map readers during the Great War. It is impossible not to be captivated by their beauty and desire to advertise their presence.

Drawing comparisons with the early evening party and the activities of these glorious insects, there is an obvious link: partying teenagers with newly acquired hormones urging them on and, congregated on the hedgerow a few hours later, these small fairies of the night leaving out green lanterns to guide their lovers home.

Forty-five years on, and the memory of the party has faded, but the sight of those sweethearts carrying their lanterns has lived with me in vivid detail.

Mammals

Brock

It is September, and Lesley and I have booked a holiday home in Witheridge on the edge of Exmoor and invited our four children and partners to join us, and of course there are two new additions in my grandchildren, Hallie and Reuben. It is a lovely barn conversion in the middle of the country, and we have our own resident barn owl living on the gable end that has deposited pellets on the floor below, allowing me to pull them apart to see what has been taken – I find the usual small bones and fur, most probably field voles.

On the first of the month, we all go for a walk, and I notice a highly active badger sett not far from the house. This is very exciting and worth returning to later to sit and watch for activity, and it is only Daniel who is keen to join me.

As dusk descends, we set off as we want to be settled before it gets too dark and ready for when the badgers will appear. The sett is located on a slope with a footpath on the higher edge with the wooded area being no more than forty metres in width. Due to the dense foliage, viewing the sett from above is impossible; therefore, we make our way to the gate at the bottom of the field which takes us across a stream and shallow muddy stretch of land that, once cleared, leads us to a spot where we have an unobstructed view of the sett. Whilst crossing the stream, I notice fresh roe deer slots that are clearly defined and recently placed.

We are now in front of the sett, sitting on plastic bags to keep ourselves dry from the damp water's edge. The sett is about thirty metres in front of us with a clear view of an opening that appears to have been used a lot with fresh mounds of earth having been worn smooth through a lot of traffic. Between us and the sett is the ever-present stream. It is a very still evening, and the wind direction is favourable so reducing the risk of being detected. The only sound present is the trickle of the stream and the occasional sheep calling from behind us, along with a song thrush attempting to usher us all indoors with his repeated beckoning.

We are sheltered by a stand of semi-mature oak, and the wood in which the sett is located is host to aged oak. Around the main opening and higher up the bank, there are badger runs with one clearly a highway across the top that leads out of the woods east and west.

Watching badgers can be a frustrating business as you must be extremely patient, not knowing whether it is going to get too dark to continue or if indeed they will appear at all. Unlike other wildlife watching, you are up against nature's clock with the risk of inevitable darkness thwarting all effort.

There! Unmistakeably, a head appears from the large opening we have been pinning our hopes on, tentative at first and then with a little more confidence. It is the boar, with his large-set jaw and powerful shoulders, and as he fully reveals himself to survey his domain, all the tension of waiting evaporates into a vapour of pure excitement and joy, always careful not to give away our position. Shortly after, two females appear with two cubs. After a day of being deep down under and disconnected from their terrestrial habitat, they take time to scope the area, ensuring it is secure. It is now approaching darkness; however, with our binoculars, we are able see clearer, and it is not long before the cubs put on a show for us by running a circuit, playing a game of tag which sometimes results in some rough and tumble.

After watching the family for forty-five minutes, one of the cubs comes up to the perimeter of their wood and takes time sniffing in our direction; we are suspects in a game of fear that has been engrained over generations. After a brief time, our presence is confirmed, and the cub rapidly disappears down the entrance of the sett, quickly followed by others.

The light has given up on us, and just as we are about to make our way home, I see a water vole enter the stream from the shallow bank opposite and swim off. More worryingly, Daniel believes he saw the shadow of a mink on the upper run of the badgers' sett.

This has been a wonderful and memorable night of wildlife watching: one of those occasions where you feel privileged to be in another creature's world. The badger remains a persecuted animal by both yobs illegally baiting them and by the establishment and some farmers with the so-called legal shooting of them to control bovine tuberculosis. Transference of TB is a two-way street; however, it is the badgers that take all the blame, and no amount of cruel persecution of the badger will eradicate the disease; this can only be achieved through better bio security and development of a bovine vaccine.

Thank you, Brock, for giving Daniel and me a small peek into your world.

Exmoor Rut

It is late October, and I have taken an impromptu day's holiday to see the red deer stags rutting on Exmoor. The unexpected day off adds to the anticipation and thrill of seeing these big, majestic animals in all their glory. I have an idea of where to look given that, three weeks ago, Lesley and I went on a half day safari with the enigmatic Johnny Kingdom who took us on a whistle-stop tour of the moors, pointing out some fine stags.

As a South Devon boy, I often visited Dartmoor which has a bleak quality that is wonderful but varies from Exmoor. Since living in Somerset, I have been to Exmoor on several occasions, and what strikes you is the juxtaposition from the soft, green coombes to the heather-clad moors, and it is the variety of the contrasting habitats that attract me. To walk in a coombe with the abundance of lush vegetation, trees in timeless tranquillity and peaceful waters rippling over smooth boulders is an experience that leaves you in complete harmony. The moors, on the other hand, create a sense of awe, and anything less than total respect will expose your human frailties.

Today I am going to Exmoor for one reason: the red deer stags, although I hope that other possibilities will open up to me. One of the things you first notice about the national park is the abundance of beech trees. These wonderful centurions not only stand in their own isolated beauty but have, over the centuries, been laid as hedges. They stand tall and arch high up, touching at midpoint across the road. Trimmers have given the trees a tight clipping to about three metres, where they are then left to express themselves. These beautiful avenues are the equal to any French road lined with Lombardy poplars. On the farms, the beech tree was first chosen for its hedging in the early nineteenth century for its quick, high growth and, when layered (known as steeping), it formed a good windbreak and strong stock barrier.

I arrive onto the moors via Exford heading towards Simonsbath with the intention of taking the first road off to the right leading onto the moor. Exford is considered the centre of the moor and is most certainly the epicentre of the hunting fraternity, and it is because of this that I feel uneasy with the place. As someone who was born and raised in the country, I do not want to see an end to country tradition, and I am not at all squeamish about the controlling of deer numbers to ensure the strongest and fittest remain to further their bloodline. My complaint with the hunt is the running-down of an animal over several miles and over a prolonged period of time, thus causing it severe stress; this is gratuitous violence that I cannot relate to.

Having taken the right directions, I now find myself on the moor, and it reveals itself in all its isolated splendour. There is a wind blowing from the north, and the day is overcast. The elements, along with the bleakness of the moor, are a perfect setting for the atmospheric rut. I eventually stop in a passing place and speculatively get out and walk across the road to a stone wall interspersed with the ever-present beech. Scanning the hill opposite, which I later learn is called hurdle down, I eventually glimpse my first stag of the day which, as it turns out, is my best. This is a twelve pointer with a beautifully shaped set of antlers that arch over his head and almost touch. This royal animal stands tall over his fifteen hinds that rest peacefully below him. Like a small child, I am awestruck and excited in equal measure; my senses stand to attention and grip me. I run off picture after picture, never getting enough. Eventually I give up the photography and just stand and admire this beautiful and impressive animal and his peaceful harem.

I decide to go back up the road and park in an area left for cars and walk down a wide drove leading onto the moor. Immediately to my right, I notice a young seven-point stag with two hinds that move warily away from me. This stag gives a poor showing as its antlers are misshapen, and one set are longer than the other. I take a track off to the left that has been trodden by the deer, and there are fresh droppings along the way. As I walk further into the moor, I unknowingly startle a flock of twenty golden plover that wheel away, giving their mournful cry. Despite the fresh signs, I do not encounter any deer and therefore decide to retrace my steps.

Back in the car, I make my way towards the coast as I can see some stags moving on the moors in the distance, and the road towards the coast will take me in that direction. When I get to the location of a stag with his hinds, I have come over the crest of a hill and have fantastic views of Porlock Bay. To my right, and silhouetted against the grey sky, is a ten pointer with ten hinds and young, and he is edgy as something is clearly irritating him, which soon becomes apparent. There is a two-year-old pricket (sometimes known as a spire) six metres away that has been seen off and is now nonchalantly, and with persistent optimism, grazing slowly towards the hinds. Having once again seen off the young pretender, the stag takes a rest amongst his hinds but only after he has announced to the world that he is prince of the moor with a deep, unmistakeable bellow which is amplified across the heather with a raised head and a clearing of the voice box.

I want to get a good picture of the stag so I start out across the heather in the direction of the boss and his hinds. Walking out to the assembled family seems like an easy task as, when you look across the moor, the heather gives it a smothered appearance and therefore something easy to cross; how wrong I am! The heather is quite deep, and the ground underneath is very uneven, on top of which there are numerous unseen trip hazards. So, I set off with all my camera equipment and finally get within a reasonable and safe distance to set up. Unnoticed by me, the persistent pricket is hidden in a hollow directly between the stag and me, and it is still bothering the big boy, who decides to get up and, with purpose and intent, walk in my direction; my initial thought is that I am the source of his ire and, after an uncomfortable moment of deliberation, I decide it would be prudent to gather myself together and retreat! It is only when I have backtracked ten metres that the youngster show itself, also coming in my direction. You must respect these big stags as to see one coming towards you, its ears pricked forward and with a menacing gait, can only mean one thing: get out of there!

Having spent a good half hour studying this stag, I move down the hill and come to another twelve pointer, but he is relaxed and showing no signs of activity even with a smaller stag moving among the hinds. Is he so relaxed because he knows his job is done and the hinds are all covered?

It is mid-afternoon and time to head home. On the way, I see an eight-point stag with well-set antlers in a field with a dozen hinds so decide to stop and watch them. However, this time it is not the deer that interest me the most but the dozens, perhaps hundreds, of chaffinches that are feeding off the beech mast. The trees are lining the road, and the finches are dropping down onto the road to feed, only to rise when an occasional car passes. Why would the birds take the risk of dropping onto the road to feed when there is so much mast in the canopy? I can only think it is because the cars rolling over the mast are breaking it open and therefore making it more accessible to the crafty little finches, thus saving them valuable energy.

I have had an enjoyable day. The stags, trees and birds, not to mention the hardy ponies, are timeless inhabitants of the moor and provide a true sense of place and certainly leave me feeling a profound sense of history and a happiness that I am blessed with a passion for nature.

Otter

Another wonderful week in Scotland, as Lesley and I always return to the same lodge in the same small community of Glenelg; it is my home from home. We are five days into our break, and four out of those five nights have been enhanced by a calling female tawny owl and, on one occasion, she was joined by her mate. I look forward to her perching in the big Scots pine opposite and calling forth, the hour does not matter, the call of the night watch more important to me.

So here I am on the 26th of March, 2014, and I have decided to rise at 5.30am with the express intention of finding an otter. The tide reached its lowest ebb an hour before, and I am hoping the rising water will bring in a bonanza of prey to draw a hungry otter into the open.

As I step from the front door, the Sound of Sleat is like a millpond, and there is not a breath of wind. The sky is overcast and, despite the mountains being dusted with snow, it is not cold. A lovely morning to go for a walk and seek out one of our most charismatic animals, and even if I should not see one, it is great to be out so early in such a magical place. There are occasions when the open country and the wildlife contained bring peace and balance to my life, and I have realised over decades of discovery that dawn is the best time to benefit from these restorative powers.

My first option is to turn to the left towards the mouth of the river as there are plenty of opportunities for an otter to fish and explore. As I set off down the road, the birds are in full voice, and the sheep are loose on the road, clattering on their tiptoes as they negotiate the noisy tarmac. There is nothing glossy or manicured about these herbivores as they go about their business in tatty overcoats and unclipped tails, and neither are they concerned by my presence.

The sound of the whispering sea on the beach is joined by calling oystercatchers as they pipe to each other. Suddenly, the quietude is shattered by an overly aggressive sheepdog that has seen me and decided I am not to be liked as he snarls and rages at my presence. Fortunately, he is penned and unable to satisfy his urge to molest me. Gathering up the scattered shards of my peace and solitude is going to take time.

No joy on this stretch, therefore I backtrack and head for the village to see if there is one in the inlet sitting in front of the pub. Again, I have no joy so, reluctantly, I decide to go further afield, which means starting up the car. Walking back to the car, I hear one of my favourite birdsongs as a curlew's soft syllables tread across the water.

Where to go? I decide to head for the ferry as there are numerous rocky inlets and shelves along the way which will offer an otter some good fishing territory. Once out of the village and onto the coast leading to the ferry, I drive very slowly, casting glances for the elusive mustelid. No luck on the long beach, so I climb the road towards the ferry, watching below to my left for any sign, when suddenly I see the arched back of something diving under the water, and my heart leaps as it is an otter. A few seconds later, and I would have missed it and passed by, such is the luck needed in spotting otters.

Parking up, I walk back to the elevated spot overlooking a collection of rocks that languish fifty metres offshore. Again, there is my otter rising, this time with a snack of shore crab. For at least ten minutes, the otter stays in the water and continuously dives, serpent-like, occasionally surfacing with a meal. This is the fifth occasion in my life I have seen an otter, but I have never seen one out of the water, so I hope my luck and patience is rewarded on this occasion.

Finally, my wish is granted, and the otter climbs among the small collection of rocks. Otters are most at home in the water, and that is clear when you watch them in their natural environment; however, it is on land that most of us get to see their physique and a bit of their character. I am sure this is a dog otter with his bulk and strong facial features. The otter is a compact animal with all the component parts built for strength and speed. The legs are short and powerful with huge paddles attached to drive it through the water, and the tail is a mass of muscle like a no-nonsense rudder to steer the animal through the strongest of currents and beyond the most difficult obstacles. The cream bib must serve a purpose that I am not familiar with, but it does show off the bristling whiskers to significant effect, and let's not forget the importance of these sensory adaptations that help the otter navigate in the murkiest conditions.

For forty minutes, I watch the wonderful animal alternate from water to rock, and on at least two occasions, it leaves spraint at the highest point; a calling card for other otters to read. The mystery of the animal is unforgettable and is reinforced when the otter decides I should go home. With no warning, and with confidence, the otter dives and does not surface again. A fitting departure and a reminder of who is in charge, and I am left with a mixture of excitement and calm, and still it is only 8am: a full day in just a few short hours.

A little over seven years later and another memorable otter encounter and less than a mile from the one described above. There is a lovely walk from the Glenelg ferry station to Ardintoul Bay that takes you through coastal woodland with elevated views of Kyle Rhea and Loch Alsh. It is early October, and there has been plenty of rain, rendering the path muddy and slippery and wetted by the burns perpetually sliding off the hill. The trees occasionally shiver excess water from their canopy to remind me of their presence.

Along the berry-rich walk, I optimistically look for otters far below where Kyle Rhea washes against rocky bays, shore and cliff face. No otters on this stretch, although I manage to see a fine bull grey seal that has spotted me and, for nearly a minute, holds my gaze before giving a muscular thrash of its tail to disappear into the depths.

Nearing the end of the path that takes Lesley and me to our destination, passing overhead a pair of raven row with strong, elegant wing beats, issuing deep, summoning and guttural contact calls. A bird suspicious of people but which, confusingly, demands to be noticed.

The weather has lightened, allowing views across Loch Alsh, and we pick our way over the pebbly beach of Ardintoul, reaching halfway between the path's end and Ardintoul Point, when in the distance I spot an otter, and I drop to the beach and beckon Lesley to do the same. It is essential to hide your profile whenever watching wildlife, and with the bank of vegetation behind us and the broken outline of the rocky beach, we should go unnoticed; let us hope the wind direction is in our favour. The otter has emerged from the most prominent burn leading down to Ardintoul Point where she has most probably been in a holt or possibly using the burn to wash away the salty water of the loch. Without warning, something incredibly special happens: two pups follow the original otter. To watch a family of otters amble down a shallow rocky burn making for the loch is a rare sight, and we have a ringside seat; please let no one innocently appear and scatter these beautiful creatures as they make their liquid descent. I have never had the privilege of watching a family of otters.

Once reaching the loch, they effortlessly slide into the water. The pups are approximately twelve months old and close to leaving the protection and guidance of their mother. One pup is evidently larger than its sibling, which would suggest the mother has raised both a dog and sow otter. The other noticeable thing about these two pups is they are fishing for themselves, which is interspersed with extended periods of boisterous play. For all the apparent independence, the family is still a unit, and the pups will follow the female ashore and are never far from her protection, not unlike older teenagers.

Slowly, the family work the shoreline looking for easy catch in the shallower waters, and as they do so, they inch ever closer to the stretch below where we lay hidden. We remain in place for over an hour, watching this fishing party working the coastline for its supper, and marvel at their liquid and graceful progress both on and offshore. Eventually, they extend their searching beyond our route home, and not wishing to disturb the family, we gingerly make our way back to the path and home.

Being aware of the needs of nature and lessening your own impact can bring huge dividends. By scouting the beach for otters and lessening our profile, this family of otters had a much-needed and undisturbed feed, and we are honoured with a wonderful view of nature at its enthralling best. Walking in the countryside aligned with nature is far more rewarding than just going 'for a walk'.

Invertebrates

Hawkers and Chasers

Mid-August and the summer has been another poor one with too much rain followed by drab, overcast days. So, when presented with a hot, sunny day, I take the opportunity to go out and look for butterflies and dragonflies, both of which love hot, sunny days. I make my way to the Shapwick reserve with a view to trying areas that I would not normally visit.

As soon as I enter the reserve, I notice to my left a scrap of untended, stringy grass with a backdrop of bramble and nettles. In amongst this grass are a host of common blue and brown argus butterflies with gatekeepers holding court among the fruiting brambles. The beautiful common blue and brown argus are small butterflies that stand out as they flit from one perch or feeding station to the next, leaving trails of colour, making it hard to focus on one to the next.

Reluctantly I step away with the promise of more riches further into the reserve. Two hundred metres on, I take the fork to the right, which leads onto the path that follows the course of the Sweet Track, an ancient thoroughfare of Neolithic man. This is a time-honoured space bathed in the gentle dappled light which breaks through the large standing oaks. To my left are reeds sheltered by alder, a peaceful space that allows for reflection and realisation that modern living is not so special. Ambling down the path, I am fully awake to all sorts of natural possibilities, when I am arrested by the sight of a silver-washed fritillary coming towards me at speed. The silver-washed is a big, strong butterfly with a confident gait and dusted in powdered gold, a real aristocrat of the butterfly world. As it disdainfully passes me by, I try following in its trail in the hope that it will alight and give me a closer look, but in the end, I have to give up and stand back to admire as it goes on its nonchalant way.

Halfway down the path, I come to a break in the tree line, allowing me to head southwards into an area of the reserve that is dominated by narrow tracks and shallow lagoons that are overgrown with lilies, yellow flag, reeds and a whole host of other aquatic plants. On the margins of these ponds are an abundance of buttercup, hogweed, purple loosestrife, hemp agrimony and even some very pretty deadly nightshade. Standing back from these marginal plants are the usual host of brambles and nettles that in turn are dominated by the mature alder, birch and oak.

The path skirting the pond is wide enough for just one person and pitches left and right through a sea of flora and fauna. There are dragonflies and butterflies in abundance and, in the brief time available to them, each pursues their own busy agenda. I slowly walk on, deciding to later return to this place for a closer look. Arriving at the end of this track, I find myself in a wider drove in dappled shade and light that is lined with oak on either side and alive with preoccupied common chasers. The males will rest on their favourite posts or branch and survey for either a mate, meal or impostor, and when they spot any one of these, they will take off at speed and give it the appropriate attention. I spend ages watching these animals, trying to understand the dynamics of their community. The other busy insect is the mosquito, and I am being eaten alive, as they are quite fearless in satisfying their hunger!

There is a small gap in the tree line that takes you further into the reserve. There is no recognisable path, just trammelled lines created by animals and opened up slightly by the more adventurous. The waterlogged ground is soggy with the black, spongy peat and indented with roe deer slots and the occasional footprint. As I travel further on, I have to take evasive action to avoid the alder branches that are trying their best to crowd me out. I am not yet ready to give up, as ahead I can hear the incessant call of a young sparrowhawk, and I want to see the source so keep going. Despite audibly getting closer, I eventually have to stop, as the going is getting tough and the undergrowth so dense that I would never see the bird even if I was up next to it.

Back onto the wide drove, I take time to observe all the activity, and my patience is rewarded with a most astonishing sight. There are honeybees feeding on late flowering bramble, and the communal good is met by their unselfish foraging. Powerfully flying down the glade comes a hornet and bulldozes into the bramble, seizing an unsuspecting bee, and once the victim has been overcome, the hornet carries its prey back to where it came from. Nature in the raw!

Retracing my steps, I make my way back to the first lagoon, as it is so full of vitality, and I want to immerse myself still further. I eventually settle on one spot about two thirds of the way down the track that is more open, and patrolling that space, there is one male brown hawker that fascinates me. This is a real bruiser, a powerhouse that dominates the pool. The brown hawker is one of our bigger dragonflies, measuring 7.5cm, which is just 1.5cm smaller than our smallest bird, but the equal of our most determined raptors, and his colouration of burnt orange and lime green flashes completes the impression of sleek hunter. The creature is so alive to his surroundings that even when a thistle seed floats upwards, he rises to it at speed, thinking it is something to attack. Eventually, the hawker's time arrives as, at ten metres above the lagoon, a gatekeeper butterfly nonchalantly makes the crossing and, without wasting any time, the accomplished predator takes to the wing and, to my surprise, ambushes the unwary. I am astonished that he would tackle something so big and yet he does so with precision, executing a swift capture and despatch. Having snared the butterfly, it is clear that the hawker is at the edge of his limit, as he struggles to carry such large prey from his aerial hunting grounds back to his lair: a branch on a nearby alder. I have watched many birds of prey hunting in this country and, like most of us, I have watched big game on the television, but watching this hunter with all his strength and power is the equal to them all.

This afternoon on the Levels is unlike any other I have spent in the past, as I normally indulge my time exploring the world of birds and mammals. What I have learned today is that if you only spend time looking for it, there are other worlds layered upon the more obvious.

Woodland Butterflies

It is late July and the first good weekend of weather we have had for weeks; in fact, this must have been one of the wettest, stormiest summers in years. They say it is due to the jet stream being much further south than is customary and therefore driving in wet weather that normally sits higher. Given this small piece of great fortune, Lesley and I head to Shapwick to look for butterflies and dragonflies in a part of the Heath that Lesley has not visited.

We enter the reserve from the west side and make our way towards the Sweet Track, an area that was inhabited by prehistoric people and where a perfectly preserved walkway has been discovered. The path we travel is tree-lined and dappled with droplets of sunshine and unusually silent given that the birds are in moult. As we come to the point where the Sweet Track is signposted, we move south through a gap in the hedge and enter a little-known secret that has Lesley captivated. This is a primordial place that is damp and lush, with tall plants growing with abandon, and all around, there are a myriad of insects. Without a cloud in sight, the sun is beating down, and the narrow path is crowded by a verdant mix of vegetation which includes meadowsweet, yellow loosestrife and great willowherb, and dotted in amongst these stoically named plants are the vetches: common and tufted. Standing tall among them are marsh thistle, with their back lit and bright red spikey stands topped off with a purple bloom and swinging bumble bees. On the perimeter, and taking a more superior view, are the stands of alder, willow, birch and oak.

The sky above the glistening pond, situated to the right of the path, is dominated by predatory brown hawkers patrolling their territories. These are the big, muscular animals of their kingdom that are watching for a mate and, equally important, hunting down other unsuspecting insects to satisfy their constant hunger. Without warning, and with great speed, they abandon the pond, and a few seconds later, a hobby sweeps through the air, low and fast, looking for a fat dragonfly to eat; there is always someone bigger in the playground to pick a fight.

Coming to the end of the path, we are upon a dappled wooded glade that has a drove running through the middle of it. Here and there, the wood has been cleared to allow dells of low brambles and wild honeysuckle, both in flower and two plants that are critical to the success of the white admiral, one of the butterflies I am here to see. The wood itself consists of tall oak and smaller birch with several mature trees and some much younger. The first butterfly to appear is a silver-washed fritillary: a large, powerful butterfly with a soft tan appearance, and I cannot contain myself as I am so excited at seeing it, and Lesley is equally impressed, especially by its colour and size. I watch one male patrol his territory, which consists of a large arc that is anchored by a mature oak with an apron of saplings, scrub and brambles. I know it is the same animal circling the tree as he has a notch in his rear right wing and is therefore easy to track. Occasionally, another sliver-washed wanders into his patch, which he vigorously pursues to either woe or chase depending on their sex, but he always returns to his circular territory.

Not long after seeing the silver-washed, I see my first white admiral butterfly. I am initially struck by the speed at which they fly and the fact that they never sit long in one place. Occasionally, they will settle on a bramble flower when I am able to admire their dapper black and white coats, and when folding their wings, there is an array of soft reddish browns on the under-wing. I notice that when the white admirals are disturbed or alarmed, they will fly high into the canopy of a tall oak tree, those great towers of life that harbour hundreds of creatures.

As well as the silver-washed having a semi-circular gash cut into its rear wing, I spot a white admiral with an identical marking. I strongly suspect they have been attacked, and I soon deduce that the culprit is one of the large dragonflies. Last year I saw a brown hawker attack a gatekeeper and kill it in mid-air, so it is possible for these two notched butterflies to have escaped the clutches of a similar attacker.

Lesley and I are enchanted by the sheer peacefulness of our surroundings and the company we are keeping. We are standing alone in a glade of dappled sunlight with semirare butterflies all around us and accompanied by common damselflies, southern hawkers and common darters. There are no mechanised sounds and no people chattering. I am completely absorbed in my environment to the point where I am bonded with it and have one of those rare moments of belonging.

We eventually and most reluctantly drag ourselves away from the dreamlike state we find ourselves in. Nature is my touchstone and has always been so. The sense of wellbeing I feel when I allow the natural world to take possession of me is indescribable other than to say it washes over me and leads to an elevated state from which so much else in my life benefits.

Mayfly

Late May and it is a warm evening and, looking out of the living room window, silhouetted against the setting sun are hundreds of mayflies. With the River Camel just a few hundred yards down the road from where I live, Lesley and I decide to drop everything and watch the spectacle up close.

Mayfly hatch in late spring and live a short life on the wing with the sole purpose of mating and setting down the next generation. This is one of nature's mass hatchings, and we are not disappointed.

The mayflies rise ever upwards like they are on tiptoe and, when reaching their upper limit, they gracefully glide back down. As they make their slow and agile descent, the males look for unattached females to mate with. Once paired, the female rests upon the surface of the river to lay her eggs. Pirouetting above the river are hundreds of mayflies, all of them using this short window to pass their genes onto the next. With the sun in front of us, the light is illuminating and highlights these delicate dots of life as they lift and descend.

Where there is such an abundant source of life there will invariably be opportunistic predators looking for an easy meal, and so it is on this occasion. Patiently loitering below the river surface are several attentive brown trout, waiting for one of our performers to get within striking distance of the river surface, whereupon they use their muscular strength to athletically rise and snatch the unsuspecting mayfly. Should a female mayfly successfully reach the river surface to lay her eggs, she is still prey to all submariners looking for an easy meal.

As if fish of various size and power are not enough to contend with, there is also a pair of grey wagtails taking advantage of the beleaguered mayfly. The grey wagtail is my favourite of the three wagtails seen in this country, as I like their pastel shading of slate grey and lemon yellow. There is also an elegance about the grey wagtail with its extra-long bobbing tail. It is mesmerising to watch these agile birds resting one minute and, without notice, rising, in a flash, to take a mayfly for its waiting chicks, snatching ever more without spilling the earlier catches. Only when they have a beak full do they return to the nest located under the bridge.

We spend an age watching this wildlife spectacle, and it is one we will not forget in a long time. The mayfly provides us with beauty, drama and tragedy, and it demonstrates that nature requires the mayfly to flourish in order that the trout and wagtail thrive. Disrupt this symbiotic balance, and all will struggle and be at risk of perishing. A lesson on how the smallest creatures benefit all, including us.

Whooper Swans

Fallow buck during the rut

Red deer in a snowstorm

Brown hares – roll with the punches

Moorland gateway

Hazy beach day

Short-eared owl

Birds

St John's Peregrines

I have recently bought J. A. Baker's book, *The Peregrine* that I have not yet started, but it has inspired me to visit the peregrines that are nesting on St John's Church in Bath. It is a sunny June day, my wedding anniversary and a perfect one for observing and photographing the birds.

When I was growing up in the 1960s and '70s and starting out on my explorative journey of birdwatching, I had a list of rare birds that I wanted to see, and the peregrine was at the very top of that list. Here is a bird that fears no one, has the grace of a ballet dancer, the raw power of a track and field athlete and the punch of a prize fighter. You don't forget your first peregrine, and you never cease to be in awe of them. Thankfully, the population has increased since I was a boy, and they have escaped the ravages of rampant overuse of pesticides, thus allowing more of us to enjoy them.

When I get to the church, I immediately see the tiercel

perched halfway up the spire and to the right of the nesting platform that has been so conveniently placed for the family to use. He is perfectly relaxed and is spending a lot of time preening. In the bright sunshine, I can clearly see the hooded, yellow-ringed eye as it constantly takes in everything around it whilst continuing its ablution. The banding of the chest and underside is striking in the summer light and beautifully defined by his dark slate back. Resting on one leg, he raises the other like a large cadmium fist and picks at it, cleaning his weapons of choice, ensuring they are in mint condition.

I can see no evidence of the young birds on the eyrie and therefore assume they are resting further back on the platform. I had, five minutes previously, been talking to a gentleman who has been following the progress of the family and who has now moved further back to try and get a view of the young birds. From behind and high up, I can hear the high-pitched screech of a peregrine and turn to look for the source of the call and see a young bird sat up on the dormer window of the building behind me and, simultaneously, the man calls to me. Overcoming my surprise, I then notice two more juveniles sat further along the roof. I don't know how long they have been there, but you must wonder who is watching who?

The young peregrine lacks its parents' majesty, and the colouration is less defined and of a rustier hue. However, behind the obvious juvenile appearance, you can clearly see the keen-eyed predator as they take an interest in everything going on around them. If a pigeon flies close by, so they bob and weave their heads as they follow its trajectory. Eventually, they move around the corner, and so

I move my vantage point to the bridge over the River Avon next to the rugby club.

As the young birds are now overlooking the river, there is a lot more gull activity, and it isn't long before they catch the attention of one irritable lesser black back gull that decides it will mob the young falcons. The birds aren't frightened as they take it upon themselves to collectively heckle the gull, which unbeknown to me catches the attention of the tiercel. From out of nowhere arrives the parent bird and sets about pursuing the gull that is now taking evasive action to avoid being caught. As it weaves with increasing rapidity, it passes in front of me and gives out a screech of fear and still the falcon relentlessly maintains its pursuit. There is no bird strike or catching of gull as the tiercel gives up the chase and pulls away, gliding effortlessly back to its perch on the steeple of St John's; he's made his point.

This is not my first observation of the peregrine, and neither will it be the last; however, it is one I will not forget. I can now add diligent parent to the list of qualities this great bird of prey possesses.

Garden Predator

An early May afternoon and I see a remarkable sight that I have not heard of or witnessed before. Low and fast, a sparrowhawk speeds into our garden and spooks the birds on our feeders, which are mostly gold and greenfinches. In their flight from danger, several birds inadvertently fly in the direction of the hawk that, despite her best efforts, is unable to snatch one of them, so she completes nature's equivalent of a handbrake turn by splaying her tail feathers, dipping her left wing and, in one smooth action, turning 180 degrees. With still enough power in her motion, the bird sets off in pursuit of the disappearing finches.

This particular sparrowhawk regularly visits our garden; she is one of last year's juveniles and is still tawny in appearance. On another unsuccessful hunting sortie, this power-packed hawk crashes into our red-leaved cherry tree. Below and behind the tree is a three-foot drystone wall, and behind that is a six-foot fence with an eighteen-inch gap between the two. The hawk drops down onto the wall and slowly stalks up and down, staring intently into the gap, looking for any prey that may have dropped and hidden there or possibly struck the fence and lay stunned. This slow and methodical search goes on for a good five minutes. Such amazing intelligence to calculate the potential reward provided by the fenced-off gap. There are many who view the sparrowhawk as a menace of the bird table, but you need to stand back from the emotional impact of their presence and admire them for their qualities of beauty, power and intellect.

This young hawk has become my equivalent of J. A. Baker's peregrines. My juvenile female shares the garden with an adult male, possibly her father: a much prettier bird. However, it is this young female that captivates me: a hawk that has made several kills in the garden, which includes two collared doves, too heavy to carry away and therefore dissected and consumed on our lawn.

One autumn afternoon, I am picking blackberries on the lane leading to the playing fields and reservoir. The weather is bright and the day crisp and beyond the bank of berries, a field that is normally inhabited by young bullocks, the same field as opposite our house, and behind me are fields of overgrown grassland. I hear her before I see her, as a deep thrust of air is pushed in my direction, doing its best to part way for the hawk driving forward. I turn to my left to see the cause of this rush, which at the same time alerts my source of interest, which in turn catches her first sight of me. In that split second, she banks to the left by no more than fifteen degrees and flies at my head height some two feet away, and we scan each other closely, she with that deep pool of black iris and vivid yellow eye giving a challenging stare with no fear attached. In one smooth and effortless action, the hawk lifts herself over the bramble hedge, drops down the other side and proceeds along the extensive hedge, hugging it close with wing tips barely missing the clipped grass.

The sparrowhawk is a wonderful example of intelligent adaptation. This a woodland bird that has recognised the benefits of moving into the urban environment. There is the ready supply of prey species plus man-made obstacles that this speedster can use to assist in its tactic of ambushing the unsuspecting. The call always goes up in our house when the sparrowhawk visits as we are eager to see it.

Tumbling Ravens

Early August and the night has passed the wet and miserable baton to the early morning, carrying it without care or shame and leaving visitors to abandon holiday plans and reach for the waterproofs. I left home early to visit Salisbury Cathedral on business, and the lashing rain dominated the journey; however, as the morning gives way to the afternoon, so the rain gives up its pursuit of the undeserving, and the sun shines with a summer breeze to soften the heat.

My journey home from this most beautiful cathedral takes me past Draycott Sleights, a wonderfully raw location with commanding views over the Somerset Levels. I had read there was the possibility for seeing chalkhill blue butterflies during warm August days on these rugged slopes. I pull up in the lay-by next to the entrance and open the car door to be greeted by a warm, bright afternoon so step out and lean against the gate to see what is on offer. To my right and climbing up into the distance is the familiar slope of carboniferous limestone, with rough grasses and wildflowers drifting amongst and either side of it.

Whilst I survey this nearside slope, I catch sight of a male blue butterfly pursuing a rival, only to be distracted by a female in her plainer apparel but just as interesting. Through my binoculars, it becomes clear that this is my intended objective and in need of closer inspection.

Just as I am about to enter the reserve and climb the steep escarpment in my slippery, best black work shoes, from the distance there comes the familiar *kronk kronk* call of more than one raven. Distracted, I decide to return to the butterflies later and instead look for these large and impressive corvids. To the north of where I am standing, and most probably over Cheddar Gorge, are nine ravens, often referred to as a constable of ravens, rising on a thermal and engaged in some aerial acrobatics. All corvids love to play in windy conditions, and along with choughs, the raven is probably the most accomplished. I have never seen so many ravens together as they tumble and drop in the wind. On numerous occasions, two birds fly in synchronised flight, for the lower bird to flip 180 degrees to reach out to the upper bird. This school of boisterous acrobats maintain their gregarious activity, heading in a northerly direction to the point where I lose sight but can still hear them.

On a dull day, all you might see is a large black bird, but on a bright day, the colour of the bird comes alive. Earlier on my visit to Salisbury, I saw mulberry trees ladened with deep violet fruit, and these iridescent birds adopt the intense hues of this bounty, and as they twist and pirouette in the reflected summer light, I can only stand and delight in their obvious joy. These dandies in their bright, shiny coats practising their moves and loudly boasting to the world accordingly.

Back on the slopes, I climb the slippery rocks, trying to obtain a footing. I find no further evidence of the chalkhill blue but do find a common blue and brown argus and decide a return visit is required.

Osprey

The alert is out that an osprey is on the Levels. It is mid-August, and this bird has visited for three years that I know of and possibly longer. It only stays for a week to ten days and stocks up on the best Somerset fish before progressing on its migration into Africa.

When I get to the Levels, it is 7pm and the weather is variable, yet warm, with a brisk but bearable wind. The evening is bright, and there loiter dense clouds a few hundred metres above the horizon but high enough to allow the sun to provide interesting and angular light. It is a crisp summer's evening that has been forged from all the elements that now compete for supremacy.

When I get to Noah's hide, situated in the middle of the Shapwick reserve, the bird is on the same postern as on previous years, which is to the south-west of the lake and situated on the far shore. This perch is a dead stump sticking six metres out of the water on top of which the bird sits and ponders. Within five minutes of arriving at the hide, this great bird stretches its long wings and launches itself into the sky.

Once in the sky, this specialist raptor gives us a fantastic flying display. The undersides of the wings are a combination of snowy white with mottled ochre on the outer edge. The upper wings are a rich chocolate brown and the under body is again white with a mottled collar. The colouration is designed to camouflage the bird from its unsuspecting prey languishing in the waters below.

For several minutes on strong, shallow wing beats, the bird circles the lake prospecting for supper and performs with a sense of ease and power. It frequently takes to higher limits, much higher sky, and uses its binocular vision as a tactic to outmanoeuvre the abundant inhabitants of the lake below. For a while, it has reached out far into the western edge of the lake, and I am concerned that I will not see it make a kill. Slowly, it starts working its way back towards the hide, giving better views, and at the same time starts to lose altitude. Suddenly, it stops all forward motion and hangs in the air like a giant kite, something worthy of its attention wallowing with equal nonchalance below the surface. Without warning, the wings fold and the osprey enters a forty-five-degree stoop, streamlined with feet tucked back, which are extended just fractions of a second before the bird hits the water. When entering the water, a huge plume is cast into the sky, lit like burnished silver by the shafts of light that strike out from under the dark clouds. The bird is completely immersed, and everyone in the hide is asking the same question: success or not? As it eventually rises from below the surface and gives itself a shake like a wet Labrador, it becomes clear that it was a miss.

Unperturbed and back into the air, the bird starts its relentless flight and once again rises to the higher levels, and it is not long before it is hovering like a small falcon before entering once more into a steep dive and again hitting the water without fear of the impact, concentrating only on the spoils. For a short while the bird cannot be seen, before struggling free of the watery world clutching its prize: a decent-size common rudd with bright sunburst orange fins radiating in the low light. The bird gives a barrelling shake of the whole body, and the excess water glistens in the radiance.

With the fish firmly attached to the undercarriage, the osprey heads for its perch, carrying the fish head forward to allow for the least wind resistance and therefore make its flight home much easier. The evening light has dropped lower and illuminates the wet bird from behind, giving a soft glow on its splayed feathers. Back at the post, the bird positions its prey and settles in for a feed.

We do not yet have these birds nesting in southern England, with the nearest being in mid-Wales and the closest English pair currently being in the Lake District. Therefore, it is always exciting to have an osprey visit us once a year and put on a show, and it is the briefness of the visit that adds to the privilege.

Short–Eared Owls

I have been a birder for nearly fifty-five years and never cease to get excited at the sight of a new bird, especially when spotted in the UK. There are a few still to be seen, and as I have never been a twitcher, the joy of finding a new bird is even more pleasurable as I have not torn up the countryside to see it; it is all about the bird, not the list. So, when I read on a website dedicated to local birds that there are short-eared owls at Chilton Moor on the Somerset Levels, I bided my time for the first sunny afternoon and set off on the short distance with Lesley to see them.

We make it to the location, which is on a piece of moorland where Chilton Drove crosses the South Drain, a large rhyne that takes excess water off the Levels. Before reaching our destination, we come upon someone with the same interest who informs us that he has found the perfect spot. We park up and get out of the car, to be greeted by a warm sun casting soft light and long shadows.

The landscape is typical of the Levels in that there are waterlogged, peaty meadows dominated by sedge. The ground is flat with the distant hills to function as a backdrop. The grasses are tall and wispy and, in places, shrouded in backlit cobwebs, and around the margins are the desiccated remains of tall teasel. Deep within the sedge, there are two roe deer, one a buck in velvet.

Within two minutes of getting out of the car and to the rear and no more than fifty metres away, sitting on a fence post is a short-eared owl, totally unconcerned by our presence. The rich golden plumage is broken up with bars of creamed butter markings. The ear tufts are clearly visible, but the most prominent features are the bright yellow eyes that, without exception, outstare everyone.

Eventually, the bird takes off, and we concentrate on the wider moor spreading eastwards, and it is not long before we see more owls. There is one roosting on a solitary tree to the north, and another appears directly in front of us, quartering the ground for field voles. Watching any owl hunting is a joy, and each has their own tactics, with the tawny and little owls preferring to perch and pounce, whereas the short-eared owl tends to use the quartering tactics of the barn owl. The owl we are watching glides across the moor and, with a flick of its wings, turns 180 degrees and scans an adjacent section. Not only is the bird using its sensitive hearing, but it is also apparent from the lowered head and concentrated look that the owl is using its bright wide yellow eyes to search out prey.

The short-eared owl is resident in the north of England and Scotland where there are an estimated two thousand breeding pairs. These owls are migrants from Eastern

Europe and Scandinavia, where anything from five thousand to fifty thousand arrive at our shores to escape harsh continental winters. The severity of the weather on the continent will determine the number of birds seeking sanctuary. The favourite habitats for these incomers are salt marsh and moorland bathed by the warm Gulf Stream where they will spend the winter. For over an hour, we watch three owls trawling the moor for rodents and never cease to marvel at their beauty.

It is getting late, and the famous Somerset Level starlings are gathering in large flocks in readiness for the full-scale murmuration, and with a song thrush repeatedly calling his parting song, we decide to head home and take to the drove. Not more than one hundred metres into our journey, we come across an owl sat on a gate post, and as we pull up and stop, it remains stationary and still does not move when I get out to take some pictures. It is only when a car approaches from the opposite direction that the bird silently glides away.

There is one final treat in waiting when we pass a field that holds hundreds, if not thousands, of lapwings. These are beautiful farmland birds and a rare sight in such numbers.

I have only one native British owl left to see which is the more elusive long-eared owl, and given the absolute pleasure its short-eared cousin has given me, I do not mind waiting as I know it will be worth it.

Gathering

The sky is clear, and as the day ages, so the westerly sky is layering more colour. With the excess of rain we have lately experienced, there has been a lack of good days to venture out and see the wildlife, especially the impressive spectacles, including the huge starling roosts on the Avalon Marshes.

Looking out of the kitchen window at lunchtime, I count nine starlings amongst the tits, finches, blackbirds, blackcaps and all. These starlings arrive with the dawn and set about aerating my lawn whilst looking for grubs and worms, and it is a mutually beneficial relationship that I am happy to encourage. To set down on this pasture, the birds rose in their millions at dawn from Ham Wall and fanned out to undertake the essential task of feeding and survival.

I gather my walking gear and birdwatching equipment and load the car. I bid the starlings farewell in the knowledge that I will see them and their kin again in three hours' time, some eight miles south-west of home.

En route and just beyond Wedmore, I enter the Somerset Levels and watch for the whooper swans reported to be in the area. It is not long before I see them to my right: a mixture of adult and adolescent birds. A rare sight in these parts where we are more used to seeing mute and migratory Bewick's swans.

I eventually arrive at the car park and make my way onto Ham Wall and head for the first viewing platform where I set up my equipment. Early sightings include shovelers, teal, widgeon, gadwall and a great white egret. It is promising to be a particularly good evening.

One of the great mysteries of birdwatching is that often when watching the more noticeable gathering of birds, you are missing something special behind you. And sure enough, after checking several times behind me, I eventually get my reward with a fantastic view of the male marsh harrier on his ghost-like flight that dips and flutters. This bird is spending plenty of time over the area where the starlings roost and is therefore most probably scavenging birds that died overnight. These will be easy pickings that are full of nutrition and will ensure that he does not go the same way as the unfortunate ones at the bottom of the reed bed.

As the sun sets, so the cold turns bitter, and it is easy to see why all animals make foraging their priority at this time of the year. As I wait for the first of the starlings, I spot a kingfisher buzzing down a rhyne, its iridescent blue back cutting through the fading light.

As the onlookers gather in increasing numbers, so small scouting groups of starlings appear but fail to settle. These birds will join bigger groups who in turn will meet up with other large groups numberings tens (if not hundreds) of thousands of starlings, when they will finally be heading for the roost. Ahead of the large groups a peregrine, without stopping, flies menacingly from west to east and looks to have his mind set on something other than the starlings.

Now the large groups have melded, and I can hear them before I see them. There is constant chattering and the sound of their wing beat. These birds harmonise in so many ways which include verbal communication as well as the ability to fly in tight formation without colliding. It is believed that the birds can fly in such carefully crafted groups and create the mesmerising shapes because they are not following the bird immediately in front but the birds several places further along, thus allowing them to react in plenty of time. We are getting one large group after another and they funnel into the reed bed, where the noise rises with birds arguing about position, with centre preferred given it will be warmer and furthest from potential predators.

Visiting the reserves on the Somerset Levels all year round teaches me so much about animal behaviour, and the rewards are tremendous. To hear the cuckoos in spring followed by the arrival of the hobbies in May, and to enjoy quiet evening walks with barn owls, hare and deer and then have a day like today is what constantly draws me back.

Ring Ouzel

Mountain blackbird, that shy symbol of wild, rugged moorland, has made an Easter visit to Somerset as it heads home from its overwintering grounds in the Mediterranean and North Africa. Enough of the stifling heat, time for home, clear air and upland clarity.

A dozen ring ouzels have descended onto the higher grounds of Somerset for two weeks in March and April 2013. We have had some unseasonably chilly weather with harsh northerly winds accompanied in places with heavy snow, and these rugged birds have battled their way through and stopped for a break before heading to the ochre and purple hue of home.

A shy bird with black plumage that, when observed at close quarters, looks as if it is wearing armour plating as a silver patina weaves in amongst the black. The same silver delineates the primaries. However, the thing that makes this bird really stand out is the white bib sitting deep on its chest which narrows across the shoulders: is it tied at the back? This is a confident member of the thrush family that bounces across the tussocks with a distinctive flicking of its tail. The confidence soon evaporates once it spots a human when it quickly takes flight.

The favoured location both for breeding and overwintering is upland hills, moors and mountains. It prefers steep crags and coombes where it can look for insects and earthworms. Once the fruit comes into full flush, then it will take advantage of this valuable additional food source.

I am at one of my favourite spots – Draycott Sleight – to see this red listed visitor, and the pasture they are browsing is the one beyond the beach trees that stand aged and proud. These great trees offer perfect camouflage to hide behind and therefore get closer to the wary birds. Although there are only a small number, it is clear they are flocking and therefore keeping each other safe from potential danger. In amongst the birds are an early arrival of wheatears that are confident and offer some additional comfort to the ring ouzels.

It is not long before the ring ouzel are spooked and, facing into the prevailing wind, lift off and, giving a few beats of the wing and banking low, allow the wind to carry them to safety. It is clear they are masters of their environment and are not intimidated by the rough terrain or the winds that scour its surface.

A great bird and a privileged first sighting.

Nightingale

This is a bird extensively and deservedly written in poetry and prose, as the song is both beautiful and mesmerising. Whether you hear the nightingale by night or day, you will never forget the moment and never cease to seek out that sound more and more.

This unremarkable brown bird could easily be overlooked, but its melodious song draws immediate attention regardless of whether you enjoy birdwatching. Aside from hearing this songster in Highnam Wood in Gloucestershire and a small patch in Somerset, all my encounters have been in Spain or Italy where they are still abundantly present.

Nightingales love untidy countryside with plenty of scrub and open woodlands with thickets and often close to waterways. They will only nest in dense vegetation that offers security and a wealth of insects. It is perhaps because of these breeding and living requirements that the British nightingale is threatened with extinction; our land and farms are victims of excessive tidiness.

Numerous times when visiting southern Spain, Lesley and I have driven down local roads and farm tracks looking for unusual birds when we have pulled up in the car and wound down the windows, switched off the engine and been absorbed by the nightingale and his song. On one such occasion, we parked the car near a scrubby waterway dominated by a stand of poplar, when the nightingale started its beautiful song and, within a brief period, a golden oriole joined in. The lyrical song of the nightingale is matched by the soft fluting call of the golden oriole and together offer up a performance in a beautiful natural theatre that washes away all cares and leaves me captivated by their concert.

My standout nightingale moment was listening to the bird in the middle of the night. We were in southern Spain near Ronda when I woke at 2.30am to hear a nightingale. I got out of bed and stood next to the open window and listened to that solitary bird as he cast forth his song. There was no artificial background noise to interrupt the beautiful sound; there was just the still, dark night, me and that bird, and it felt emotional. A few minutes into the rendition and a distant scops owl offered its rhythmic base to the orchestral delicacy of the nightingale. In the inky darkness and without the benefit of sight, the only alert sense is hearing, and the nightingale's song was the unforgettable focus.

In my quest to hear a nightingale in the UK, Lesley and I head to Highnam Wood in Gloucestershire which is reputed to have a strong community of these melodic vocalists. When we arrive at 8.30pm, I am surprised there is only one other car in the area and the car park locked.

We choose to walk the circular walk that circumnavigates the coppiced section that is full of deep, impenetrable scrub: ideal nightingale habitat. It is a beautiful, clear evening, and the wind is low, allowing the birdsong to dominate, and the most penetrating of those is the song thrush of which there are several birds calling at full volume.

At Highnam, there are the bluebells and young bracken and ferns pushing through. There is greater stitchwort, white dead nettle, germander speedwell and one of my favourites – pink campion – which I look for every year. The composition of this wood is quite different to anything I am familiar with. To the perimeter of the wood are great stands of oak and ash which are separated from the coppiced section by wide avenues or rides.

We decide to take a detour and climb one of these rides which supposedly has a magnificent view that we manage to miss. As we continue to walk the path, so the wood closes in and the dimpsey light turns to an encroaching darkness with tall, interlocked pines leaning over us. We turn around and head back and find ourselves entering open space and increasing light. On the way back to the coppiced wood, I spot the looping flight of a woodpecker which lands on a welcoming bough of a tall oak and shuffles its length. It is immediately apparent that this bird is too small to be a great spotted woodpecker but is in fact its smaller, much rarer, cousin, the lesser spotted woodpecker. A first for me which demands more luck than skill to spot.

As darkness takes control of the proceedings, so the cacophony quietens, and the daytime inhabitants settle for the night. As dusk ambles out of the wood, so darkness appears in a hurry, and we enter a glade with the copse to our right. As we stand still for a moment, so the nightingale starts his beautiful liquid song that lifts my spirit. There are other birdsongs I love to hear, but there are very few as addictive as this.

It is time to leave the wood as the vocalist has gone quiet, and I believe we would have to remain here well into the night to hear more of this chorister. On the way home, the full moon appears cloaked in orange and rising from a bed of mist.

Golden Eagle

It is a bright and cold early October day, just two days after my fifty-eighth birthday, and I am embarking on one of my favourite walks, if not my favourite place of all: Glen Licht. The sky is blue with picture-book clouds, blown and bright. There are dustings of sunlight gracefully brushing the skirts of the retiring five sisters. A strong easterly blows down the glen with a raw edge to its strength, delivering a sharp reminder as it marches over and past Lesley and me.

The River Croe, exuberant, journeys into the fallow months, swollen with fresh mountain water that bumps and jostles towards Loch Duich. The shimmering shingle banks of lost summers are submerged, and the nesting pair of oystercatchers have seen their young take flight and they, in turn, race the waters to the coast. Something substantial rises from the water – is it a silky brown trout or the elusive salmon, prince of this imperious land?

Finding a sheltered spot next to the Croe, and in the presence of a rowan bursting with blood-red berries, we sit and have a break of coffee and cake. The tree is beautiful in its finery and defiantly leans out over the river; a dare in conviction it will win. The rowan is everywhere; an age-old superstition determines planting one near your house ensures a happy home, however it is bad luck to fell one. Seated on some rocks and behind us, we can hear a chuckling burn, rolling over rocks and rust-coloured granite pebbles, a happy sound.

Onwards, as we are only halfway down the glen, and as we turn the corner, we are again met by the relentless easterly, but it doesn't concern me as the views and walking are wonderful. All around, the stags in the mountains vocally jostle, and high up on Meall an Fhuarain Mhoir I can see three stags in close proximity, barely tolerating each other but each with his own harem to defend. This is a restless time for these big animals, and for some, the exertions will take their toll over the upcoming challenging months.

Finally, we reach the end of the glen where the university house stands alone and barred; fortunately, we are not looking for shelter, although somewhere to have lunch on the leeward side of a wall is welcomed, and so we crouch low behind the neighbouring derelict building. We are alone in this beautiful place enjoying lunch when I catch the unmistakable sight of a golden eagle that, unperturbed by the driving wind, drops down on outstretched wings from the top of Meall an Fhuarain Mhoir, harnessing the power of the wind to avoid expending too much energy. I wonder if she has seen us; of course she has, with her powerful vision, but this doesn't seem to stop her from drifting lower into the glen, giving us superb views, and as she glides from right to left, we get the most amazing insight of her effortless power. Just as I think she is going to disappear over the last of the Five Sisters, she dips a wing and effortlessly glides across the mountain face, clearly intending to flush some prey. Time stands still as this majestic bird, without a single beat of the wing, sweeps the mountain surface, and still she keeps up her vigil, and I become convinced she will land. Momentarily, I lose sight behind a rock and, despite desperate searching, she is gone. Then Lesley calls that she is high above the mountain – how did that happen, what force drove her three thousand feet in such a small blink of the eye? She climbs higher and higher until only a dot in the sky and the whole experience will stay with me forever.

This is not the first time I have seen golden eagles in this wonderful glen. I have seen one at altitude wheel out of the mist and circle a calling stag. I have watched with Daniel a pair in spring, wings folded, rapidly swoop at forty-five degrees across the glen to land on Meall an Fhuarain Mhoir and call loud and proud that unmistakable sound of the wild. And with Fiona and Lauren, watch the same pair two days later wheel lazily in a loosened bond.

As we head back down the glen, I am walking on air – nothing can tire me, and nothing can steal that memory or all the other eagle memories this wonderful glen showers upon me. The sun is low, and we are close to journey's end when a shard of light picks out a low-spread juniper bush, a blast of vibrant green on the bracken face of on Meall an Fhuarain Mhoir.

Loch Duich and the five sisters

Peregrine falcon

Roughtor, a moorland view

Isle of Skye looking west

Thurlestone Rock

Female otter and two cubs

Bait Ball

Lesley and I are visiting my mother and David at Thurlestone for a few days and then onto St Ives for a small break. Prior to arriving, I have been reviewing the local ornithological society website which has records of yellowhammers in the area, a canary-like farmland bird associated with my childhood and so sadly rare today! So, this morning – Monday the 14th of January, 2019 – I decide to seek them out. I leave the house and head west towards the sea with a view to checking adjacent fields. The day is overcast and the sea flat, calm.

After a short distance, I scan the sea for any activity, hoping I might see something unusual. Approximately a mile out to sea, there is a lot of activity with many birds and mammals involved. I have a great set of binoculars, but picking out the detail is proving difficult, and I notice in the car park someone with a 'scope looking at the same scene. Time to backtrack and introduce myself and get an understanding of what exactly is happening in Bigbury bay. It turns out that the gentleman is Mike, and he has a wonderful view of the action.

Mike kindly allows me look through his 'scope, and what I see is remarkable and like something from a BBC natural history programme. There is evidence of a feeding frenzy, and the variety of creatures involved is significant, and in contrast to the surrounding sea, the churned water froths with activity. Piecing together the evidence, it is probable a pod of common dolphin has pushed a shoal of fish, possibly herring, to the surface and created a bait ball. These unsuspecting fish then became easy pickings for all, which includes approximately thirty dolphin, diving gannets, red- and black-throated divers and various gulls, on the surface as well as circling from above. Slightly to the left of the gathering are fifteen razorbills picking off stragglers. As the drama unfolds, so more players come on stage, with dolphins arriving from afar, accompanied by scores of birds, and the audience of two onshore have the spectacle to themselves.

I am so excited I call Lesley to join me, and Mike adjusts his 'scope to give Lesley the chance to view the action, and she is equally mesmerised.

My wildlife experience has one more surprise. Entering stage left arrive eleven common scoter, flying low and fast past the feeding frenzy and landing approximately four hundred metres away. Accompanying the scoter is a male long-tailed duck. Common scoter and long-tailed duck do not hunt fish, preferring shellfish, crab, small fish and vegetable matter, so these birds nonchalantly ignore the surrounding excitement, perhaps picking up scraps and small escapees.

After watching for one hour, we drag ourselves away, like leaving an impressive performance before it has finished. I am fortunate to have seen some wonderful wildlife spectacles, and this ranks with the best as it is not every day you witness a multi-predator feeding frenzy and for so long. Not only was this spectacle a first and probably only chance for me, but I also observed other firsts such as common scoter, long-tailed duck and overwintering razorbill.

I feel so privileged and humble to have been in the presence of such a special wildlife spectacle. Nature matches, and sometimes trumps, everything man has to offer, and if only more took the opportunity to embrace the beauty and grace presented by the natural world, then our beautiful blue planet would not be in such a mess as it is today.

Call of the Wild

Over the last two years, I have volunteered with the RSPB to help carry out a breeding wader survey on the Somerset Levels and Moors. Having first joined the RSPB when I was ten, subscribing to the Young Ornithologists Club (YOC) and been a member for over fifty years and now semi-retired, I have decided to give something more than a subscription.

The breeding wader survey has been ongoing for decades with excellent historical data to make year on year comparison. There are four birds of interest: redshank, curlew, snipe and lapwing. All these birds should be breeding abundantly on the moors and yet their number have been in freefall; therefore, it is important to understand where the last remaining pockets are located and what we can learn from them, how to protect them and, hopefully, expand their territory. The post-war extensive and efficient draining of the moors accompanied by higher cattle density, plus switch from hay to silage, have all contributed to the steep decline. Further impact to numbers has been the more recent decline of invertebrates required to sustain young chicks. Predation too plays its part in the disappearance of these birds.

This year, I and two others have been allocated two areas to survey with the requirement that we visit each site over three months, and these are Tealham and Tadham Moors and King's Sedgemoor. Our first visit to King's Sedgemoor is on the 22nd of April and to Tealham/Tadham on the 26th of April, 2019.

We arrive at King's Sedgemoor east, 'Site of Special Scientific Interest' (SSSI) at 6.45am, which is a little late but not a problem given that it is April, and the sun is rising later in the day. We park at the drove near Greylake village and are drawn into the survey area by the rhythmic call of the cuckoo, the iconic announcer of spring.

Aside from the cuckoo, the first thing I hear is a curlew. The slow melancholic call of the wild, building in depth and urgency and carrying me back in time to my childhood. This is a sound I never tire of, and indeed it is one of the treasured few birdsongs that I will enthuse with anyone, including members of the bird club I run. In the first years of my birding life when living in Kingsbridge, I would wait for the winter tide to recede, exposing the mud flats, when I would walk to Bowcombe Creek, following the estuary with my new binoculars, looking for the overwintering waders with one of the favourites being the curlew with its purposeful gait, long probing bill and call from the depths of time. How could I have known then, in the late sixties, that this iconic bird that seemed so numerous would be catalogued as endangered on the 'Birds of Conservation Concern' Red List, a troubling and sad fact! So here I am on the Somerset Moors, clipboard and pencil in hand listening to, and watching, these wonderful bird as they prospect for a suitable nest site, what a treat! Ghosting in amongst the birds and shrouded in a light mist are hare, close enough to watch but distant enough to not bolt. Roe deer in large numbers, broken up into small groups, watch me intently, and they are not so blasé as they trot then run for cover.

As I walk each field looking for more birds and potential nests, I flush two snipe. This is a bird that will sit tight in the hope that you will pass it by and, when it loses its nerve, shoot out of its hiding place at breakneck speed and make you jump. This is another bird to remind me of my youth when, as a teenage boy living at Thurlestone Rock, the marsh outside my bedroom window would flood and with it there would be an influx of waders and wildfowl, and probing amongst them would be the wary snipe. A few days later, on Tealham Moor, I hear numerous drumming snipe (see the chapter called 'Drumming').

We stay on King's Sedgemoor for two and half hours when, sadly, I must drag myself away. On the way home, we stop off at RSPB Greylake to take in the sight of a reported yellow wagtail which shows beautifully.

Onto Tealham and Tadham where the scenery is breathtaking in its open desolation, the big sky beckoning me on and the rising sun scuffing up a mist from the lethargic rhynes. On the moor to the north lies the best habitat for wading birds and in the damp meadow are swathes of marsh marigold and lady's smock, plus many other plants and small obscure flowers that would enthuse a more accomplished botanist. I take the drove running alongside the Bounds Rhyne, rutted with deep peaty tracks and tall wet grasses. Halfway along the track and to my right, I hear a chipping snipe which is a great indicator that snipe are still present on this moor. When I finally get off the track and back onto Totney Drove, my trousers and boots are soaking, and therefore it's good to be back on the road and heading west.

Where the drove takes a sharp turn to the right, I see two curlew-like birds feeding in a cattle field. I watch them for a good ten minutes when they take off with a piping call that identifies them as whimbrel, on passage to their northern breeding grounds, an unexpected and pleasant find. Back down the drove and off to the right, I walk the track to the River Brue where, sadly, the adjacent fields are intensively managed for silage, meaning there is little habitable or wader-friendly land or food. I walk the margin of the river, going east until I pick up Jack's Drove and head back to the main road of Totney Drove. Halfway along Jack's Drove, I am surprised when I see a quartering short-eared owl; this is a bird associated with winter, not as we approach May. I am joined by the others doing the survey with me, and we spend a good ten minutes watching the owl. Late in the survey, before leaving for home, two curlew fly in from the south, calling as they arrive and land in Tealham.

Such important work for the RSPB as hopefully we can stabilise and improve breeding wader numbers and great that I can discover new habitat with such wonderful sights.

It is hard to do justice in words the mystery and magic of this place that has traversed time. Flat, open land stretching into the distance, broken up by rhynes that steam in the early morning sun, trickling mist creeping across the damp marshes and towering skies beckoning the eye to infinity. Cuckoo calling, hare and roe deer freely claiming the moor as their own; millennia of occupation provide them with certainty. The curlew's melancholic call anchored to the moor and snipe carving a dash through the still air. These visits have given me the chance to immerse myself in the present, splendid isolation and reflected history of the moor and to revisit iconic birds and animals of my youth.

Drumming

A Sunday evening in April, and I am restless to hear the calling of a wild bird of the moor. I am tired, and it would be easy to slouch in front of the television, but hey, carpe diem! So, 8pm I set off for Tealham Moor, and thirty minutes later I park and walk Totney Drove in the hope of hearing snipe drumming in the evening sky. I know snipe are present as, two days previously, when on an RSPB survey, I heard one chipping.

A short distance up the drove, a curlew arcs slowly in front of me whilst delivering its beautiful and melancholic call. Landing in the field to my right, I unsuccessfully search for it when it gives a couple of minutes of contact calling to a nearby and hidden mate. Having indulged myself with some curlew infatuation, I turn back and head towards a location where I have the best chance of hearing snipe.

Snipe have two distinctive calls: chipping and drumming. The first is a continuous chip call said to be made by both sexes when stationed on the ground and the 'drumming' made by the male in flight during his courtship and territorial display. I want to hear both calls, but I am especially drawn to hearing the male with his distinctive drumming. The sound is rhythmic and mesmerising and would not be lost in a professional orchestra, and the birds' ability to use the moorland acoustics to cast the sound far and wide must be astounding. Seeing a snipe is virtually impossible as the various shades of brown patterned on its back blend perfectly with its habitat, and if you should disturb one, it will discharge itself from under your feet like an Olympic sprinter.

The sun sets at 8.25pm, so it is now getting progressively darker, and with the overcast sky, the process is exaggerated. It is now 9pm and, under a dimpsey sky, standing at the corner of Jack's Drove and Totney Drove, I hear my first chipping bird, and its persistence lasts for several minutes. The response is not long coming as a male rises into the sky to start his drumming, and once he has met the pinnacle of the climb, he pitches himself into a roller coaster of dives and ascents. On the downwards journey, he splays at right angle the two outer tail feathers which reverberate with the passing air and produce a wonderful fast drumming sound that rebounds across the moor. The chipping increases in intensity, leading me to believe it is a responsive female.

The gauntlet is down, and other males rise to meet the challenger and set their own note within the boundaries of their territory. Soon there are five birds that I can hear and more possibly further afield, all riding the skies, intent on defending their claimed piece of moor. It is now virtually dark, and it sounds like the drumming is more prevalent further up the drove and therefore time for me to get closer to these musical masters. When I have got to my new spot, it is clear I am under the action as there is a nearby female chipping, and a male on his downwards stoop sounds as if he is going to land on my head before his drumming quickly recedes, only for the process to be repeated in the dark. Nearby, another male is performing his own score and, once attuned, I obtain a good impression of how close they are to each other. I stand for over twenty minutes listening to these secretive birds shrug off their inhibition and, unabashed, proclaim their presence. Eventually, one by one, these competing birds refrain and resume their earth-bound concealment.

Listening to birds in the dark is magical. It is not a common experience and is often limited to hearing owls; however, being in a location before and during the dawn chorus or standing close to a nightingale as it serenades through the night are wonders not to be missed. Amongst these great moments, I can now add listening to snipe over a remote moor, drumming in the dark, and out here on my own in this vast, timeless space, it feels like time has stood still for millennia, an unforgettable experience that these reclusive birds have gifted me.

Whooper Swans

Whooper swan, a bird of grace and dignity, a distant traveller imbued with mythological stories, borne on broad wings, white like pressed linen. An elegant and loyal family bird that travels from one distant land to another and on each return fills me with anticipated joy.

October 2020 and I am staying in Glenelg, West Highlands of Scotland, and Lesley and I agree to go for a walk along the road to Arnisdale with our friend Jeannette who has promised to take us to a secret bench with magnificent views. The weather is changeable, and when each squall of rain is washed away on the stiff northerly breeze, the clear air offers expansive views. It is quite a slog up the hill and, having walked two miles, we enter some sheltering pine trees, and when emerging from this small stock of Sitka spruce, we stand high on the edge of the Sound of Sleat with a precipitous drop to the sea below. Opposite is the Isle of Skye; to the left are the Small Isles of Canna, Rum, Eigg and Muck. To the right and looking north are the distant mountains of Torridon. Our vantage point is second to none and the views stunning.

We settle on the bench and break open our flasks of tea and coffee and take in what nature has to offer. Far below, a buzzard works a thermal and, with ease, rises upwards and beyond; how often are you elevated above this expert of the shifting air and able to watch it plough its unseen furrow? Gannets plot an unbending course down the Sound, always watching for something to plunge upon. Two days earlier, I had watched a pair of porpoises working up this very channel and in the past seen bottlenose dolphins pirouetting and skipping down the waters, and I am ever hopeful we might see more cetacean. Skywards, something captures my attention, and my birding instinct tells me it is special; training my binoculars on it reveals a white-tailed eagle forlornly pursued by a grumpy pair of puny-looking crows.

The weather is threatening to break so, having enjoyed thirty minutes at our elevated vantage point, we set off home but first shelter in the spruce as the rain is bouncing! The return journey downhill is proving a lot easier. Where the road levels off near the River Morar, I stop to take one more look at the Sound and there, flying from north to south closest to the far shore, are approximately thirty birds in a V wedge. Lifting my binoculars, I expect to see a skein of geese but am happily proved wrong as they are whooper swans, and I can barely contain my excitement. This wedge of swans have, without a break, migrated from Iceland and, through night, day and unpredictable weather, completed the six-hundred-mile journey, maintaining verbal contact throughout. Unrelenting, they maintain the pace, with the lead bird forging a slipstream for others to fold into, and once the leader loses strength, another takes control, and the tired bird settles further back. Teamwork towards a common purpose.

There are many myths, stories and fables attached to the whooper swan which can only be rivalled by a small number of other species. One of my favourites is the swans returning north in the spring taking the souls of the dead with them to set them free. I was so heartened by this story that I sculpted a swan's wing and placed the words on the underside to remember my grandson Luca when he was tragically taken from us at such a young age.

The native mute swan rightly endears itself to the nation, but for me it is surpassed by its large cousin from the north, transporting myth and mystery and, through the cold, dark winter months, offering hope of brighter, warmer days to come. To stand and watch whooper swans in the stillness of winter is life-affirming.

Pentire Choughs

July 2022 and we have been enjoying a sweltering summer; therefore, Lesley and I decide to introduce our friends Tony and Barbara to one of our favourite walks, the Rumps, Pentire Point and Pentireglaze.

We set off early in the hope of avoiding the peak of the heat and park up near the Pentireglaze mine which, during the nineteenth century, mined hundreds of tons of lead and a reasonable amount of silver. Adjacent to the car park and pitching on a chorus of song is a skylark, reluctant to rise too high, showering us with a cascade of notes.

The first thing that strikes you when reaching the coastal path are the various shades of blue and turquoise. The complementary sea and sky compete to offer the most striking spectacle of blues, turquoise and a distant horizon of tinted pink. The sky is awash with blues leading dark to light, offset with thinning and loose brush strokes of cirrus cloud that are hopelessly outcompeted by the heating sun.

The sea is more complex in its choice of palette, featuring turquoise, ultramarine and a distant use of cerulean, and when you look directly below the surface, the clarity takes you to the very depths. And staging this beautiful sight are the bold up-reaching angular cliffs and secluded coves that lead into the distance as far as Tintagel.

Not long into the walk and I hear the unmistakable call of a peregrine. To my right and above Carnweather Point is the falcon and one of her young, and it is the youngster that is making all the noise. The young peregrines are now fledging, and their survival skills need establishing and perfecting, which include high-speed flying lessons and agile prey hunting. I witness the falcon flying in with a kill and the young bird arcing upwards to meet its mother, and as the noisy bird gets close, the female drops the prey, and the young bird flips onto its back and, with outstretched talons, catches its breakfast. It is one of those occasions you read about and hope to see, and having witnessed it, I can understand why such moments are cherished; it was spectacular. Now I know we are in for a great walk.

As we walk the undulating path, we have excellent views of a kestrel, a coastal specialist, wind hovering and scouting for prey. Climbing upwards, we see a rose-kissed cock linnet perched on the upper reaches of a gorse bush. Normally linnets are shy and will scatter when approached, but this beautiful bird is completely unconcerned by us and continues singing its dainty and enchanting song; indeed we get so close you can see with the naked eye the delicate colours of its plumage.

Reaching Com Head reveals a sweeping panoramic view of the coast to include the Rumps and Mouls island.

The distant and subdued colours of the cliffs offset beautifully against the showy sea and sky; however, this is deceiving, as getting closer will reveal a depth of colour and variety of plants in the landscape. The leading line of the path beckons, and so we continue but never lose sight of the breathtaking scenery.

Walking towards the Rumps reveals plenty of summer plants which in turn support a plethora of butterflies. The walk and open land before the Rumps are predominantly grassland, awash with the umbels of sea carrot resulting in a snowy landscape set in the July sunshine. The most numerous butterflies are the meadow brown enjoying the abundant food source provided for its caterpillars in this large expanse of grassland. There is a diverse selection of plants in this unimproved land, allowing for a wide variety of butterflies and moths to thrive. The larvae of butterflies and moths are mostly limited to specific plants on which to feed, and without the plant of choice, the butterflies will not breed and thrive.

As well as a variety of wild grasses and sea carrot, there are many other plants and flowers which include bugle, bird's-foot trefoil, knapweed, charlock, wild thyme and lady's bedstraw. And all around there are the desiccated remains of the spring blossoms such as thrift and sea campion. With these and other flowering plants come the accompanying butterflies and moths which include red admirals, large white, wall brown and six-spot burnet; however, my favourite is the dark green fritillary. It is always a great moment when you see a fritillary, and this is no exception as I see at least two individuals power their way across the landscape on their broad wings.

Three of us walk out to the headland of the Rumps whilst Tony decides to walk up the adjacent outcrop. I know that puffin, guillemot and razorbill nest on the Mouls; however, I cannot spot any today as they are too far away or fishing on the opposite side of the island. With careful steps, we round the headland and join Tony to continue our walk, stopping at a convenient bench to quench our thirst and enjoy a piece of cake.

Approaching Pentire Point, I take the opportunity to look down the cliff face and observe the sea life. Given that the sea is calm, it is easy to pick out a moon jellyfish as it appears to aimlessly drift, and a little further along, I see a large bull grey seal that is as curious of me as I am of him. Scanning the seas does not reveal any cetaceans or basking sharks; that is one for another day.

Eventually, we reach Pentire Point, a rugged sentry of rock, and peer across to Padstow Bay, looking towards Stepper Point and the towering and clearly visible Daymark. Continuing along the path, we can now see Polzeath and the multitudes enjoying the beach and sea.

My senses are finely tuned to birds in their environment and, specifically, I know what to listen for and where, and Cornish cliffs hold many possibilities, with the biggest prize being the chough. Sure enough, I hear the unmistakable *chack, chack* of choughs, and suddenly there are eight to ten birds around us: parents and fledglings. These are wonderful birds in their loose-fitting black overcoats, bright red stockinged feet and curved red beaks, perfect probes for grubs in the short clifftop turf. Chough in flight is a thing of joy as they bounce and ride the swirls of energy offered up by the coastal cliffs. This is an exciting moment of the walk, so exciting that I lose my sunglasses!

This wonderful walk never fails to deliver, and today has been no exception. Good company, kind weather, stunning views and remarkable wildlife. To have the walk book-ended with peregrine and chough is the perfect start and finish to a memorable day out.

Locations

Levels Walk

It is early evening on midsummer's day with the sun perched low in the cloudless sky and casting a soft orange glow across the Levels. There is no wind, and the wildlife is quietly going about its end-of-day business. I have parked on the Shapwick road next to the nature reserve, and Lesley and I are undecided on whether to cross the road and enter the reserve to the west or walk into the main body of Shapwick. Our mind is made up for us when we hear a family of little owls calling from behind the hillock of excavated peat. We want a leisurely walk as Lesley is not long out of hospital and gently working her way back to an acceptable level of fitness.

Having crossed the road, we pass through the gate in the direction of the setting sun. We look for the owls but no joy as, once we come onto the scene, they fall silent.

With the large rhyne known as the South Drain to our right, we walk along the rutted path accompanied by the lush reeds pushing and shoving to get closer to the water's edge like eager bathers waiting for the first to take a dip. The water stands still with a multitude of lilies basking in the sun and scattering their tight-budded yellow flowers in trails as far as the eye can see. There is enough warmth in the air to encourage a respectable number of small tortoiseshells to still be skipping from one bloom to the next.

We are privileged in that we have this corner of the Levels to ourselves; indeed, it feels as if we have the whole of the Somerset Levels to ourselves. We are connected; an invisible bond between us and this quiet, aged land. Beyond the reeds lies Westhay Heath, a section of moorland that is pockmarked with clumps of sedge and bordered by willow that were once pollarded and harvested but whose withies now show their own age.

We eventually reach the bridge that straddles the Drain and come across a five-bar gate that prevents us from entering Catcott Heath, a land of tall scrub and sedge, an undisturbed wildlife paradise that looks hostile to the untrained eye. Directly in front of us, barn owl and kestrel families inhabit two owl boxes, each on sturdy poles.

Passing through the landscape are ancient droves, and to the south of the Drain, and within Catcott Heath, lays one such passage called Shepherd's Drove. These droves were used by shepherds for driving their livestock, normally sheep, from one location to another. As we stand surveying the landscape in the hope of seeing an early evening owl, we catch sight of two roe deer running down Shepherd's Drove east to west, one chasing the other. Given that the roe deer are coming into rut, my first thought is that it is a buck chasing a doe; however, once I have focused my binoculars onto the pair, it becomes clear that it is one buck chasing another over a territorial dispute. Both deer run down the drove towards a small grove of alder and birch that is bordered by a fence. Once at the grove and hemmed in by the obstacle, the lead deer is confronted with the option of turning to fight or taking the only available exit route: jump into the rhyne. With an ungainly leap and almighty splash, the buck chooses the second option and takes flight across the water and swims for the far side, where it hauls itself out and, accompanied by its dented pride, sets off to try its luck elsewhere.

Excited by this unexpected and dramatic encounter, it takes us a while to settle back into the slow rhythm of the peaceful evening. I continue to search out my owl whilst Lesley turns and looks back for anything of interest behind us. Interested by rustling from within the reeds on the far side of the rhyne, Lesley spots the unmistakable black- and-white headgear of a badger who is actively seeking a meal from the base of the reeds where they meet the water's edge. This is a big solitary animal that is out unusually early, and I can only think it is a displaced boar.

This is turning out to be a wonderful walk that is full of surprises. The original purpose of the walk – a gentle evening stroll in the country – has been overlaid with rich layers of encounter and discovery. There is one more dramatic turn awaiting us, which rivals the duelling bucks for its impact.

As we resign to leave with no sign of the elusive owl, we prepare to depart in the same direction as we arrived. To the left of us there is the broken hedge of hawthorn and to our right some scrub, stunted trees and a large willow. Suddenly, from the left, passing low and fast through the hawthorn, flies a finch, wheeling and banking, and it is soon clear to see what it is trying to shake off, as in hot pursuit, and matching every twist and turn, is a merlin. The birds are level with our eyeline, and Lesley instinctively points and asks, "What's that?", and it is that momentary reaction that breaks the bond of hunter and prey as the finch peels off to the right and the merlin crashes through the low tree to its left. A merlin is an amazing bird; it is our smallest falcon, packing a punch above its weight, and to even see a merlin is a joy, but to see one hunting with such determination just a few feet in front of us is a privilege.

The Somerset Levels are special as there is a wildness about them that is set in the most tranquil of surroundings. There are many places around the world that can equal the Levels, but there are few that can better them, and this evening has provided us with plenty of action and rare sights.

Dawn Chorus at Black Rock

It is 5am on a May morning and getting lighter by the minute; time to get out of the car and pull on my boots. I have parked on the roadside at the top of Cheddar Gorge with the aim of walking through the woods and onto Long Wood via Black Rock and, more specifically, to hear the dawn chorus at its peak fitness.

As I set off, the calling birds are in full swing. The first voice that calls to me is that of a tawny owl that gets a response from two neighbours. Paradoxically, these owls are calling goodnight and laying down audible boundaries whilst all around them others welcome in the day. We have only just come out of a harsh winter which always hits owls the hardest as the snow carpets the landscape, thus reducing the available food source. It is therefore heartening to hear them in full voice.

Entering the reserve, there are pockets of lesser celandine, and the red campion is looking vibrant in its fresh showing. A flower that I have not come across before is the yellow archangel, lemon-yellow flowers on tall stems of nettle-like leaves. Above this colourful display is a fresh lime-green canopy that is still transparent enough to allow sufficient light through.

The most assertive voices that carry through the wood are that of the song thrush, blackbirds and diminutive wren, a bird that is so eloquently captured by Ted Hughes in his poem 'Wren'. Each of these birds has a charm and special relationship with the British public. I love to hear and see all three birds, and if asked to vote a bird of Britain, then I would choose the wren over all others, including the robin. The wren is a pugnacious little bird and the only songbird to be represented on a British coin; I can remember as a boy going through a jar of my grandmother's wren farthings.

The song thrush has seen a big decline in its numbers over the last few decades, which sadly is down to the persistent use of pesticides that kill its main food source: that of gastropods. The sound of the song thrush is a voice that lightens the heart, a creature that recognises the decline of winter long before others, choosing a high perch with good acoustics to repeat its presence. Thomas Hardy eloquently records the joy in his poem 'Darkling Thrush', although some argue this is a reference to the mistle thrush.

My amble through the woods – it could hardly be called a walk – takes me deeper into the amphitheatre and, with slow deliberation, allows me to become absorbed in the impressive performance laid on for me. With no one else present other than my feathered companions, I am assailed by their song as they cast, full throated, in my direction, determined that male competitors keep their distance and to reassure partners of their presence. Entering the depths of the wood, I am stopped short by a beautiful woodland song – that of the redstart – although, sadly, I do not see it. The sound trickles in my direction like a summer stream skipping over a smooth bed of pebbles. I am transfixed by the beauty of the sound and stay to enjoy it for a good ten minutes without any real need to look for the colourful soloist.

Eventually, I find the will to move on and make my way towards Black Rock. Coming out of the wood into the open, there is the distinctive punchy and blunt call of a raven. Over the centuries, these big birds have had a chequered reputation, being mostly treated with disdain and disliked within the farming community because of their perceived inclination to predate young livestock, but they are also revered as guardians of our safety whilst resident at the Tower of London. Ravens have one of the largest vocabularies in the bird world, and watching their tumbling aerobatics is an immense joy.

Walking with a bit more purpose, I leave Velvet Bottom to my right and head towards Long Wood. When I get to the junction that leads either to Long Wood or up onto the meadows of the high ground, I am tempted to take the left fork and head for the hills as I know there is a community of skylarks resident there. After short deliberation, I stick to my intended course and pass through the gate into Long Wood, and I am not disappointed. The woods are bathed in translucent light, and the newly appearing green canopy of beech leaves is offset by the most beautiful showing of bluebells. I take a fork up to the right of the main footpath which leads into the heart of the wood, and the perfume of the blooms is heady and intoxicating. The wood is echoing with birdsong, and the only sign of mammalian life forms are the lines trammelled through the bluebells, footpaths laid down by animals that choose to live outside of our boundaries.

Reluctantly, it is time to head home so, slowly, I retrace my steps, always watching and listening for the unexpected and, just as importantly, the familiar. As I arrive at the location of the redstart, he is still calling, and I wonder if he saves his special song for my arrival?

A walk at dawn is one of the most magical experiences you can undertake, and on one level, it is a shame that more people do not make the effort; however, on another, more selfish, level I am glad they do not. To momentarily escape into the crisp beginnings of a world that is inhabited by the multi-layered beauty of nature and devoid of all things peopled gives life a deeper meaning and leaves you feeling humbled and at peace.

Mendip Dawn Chorus Walk

Time for my annual dawn chorus walk, and this year I am joined by Sam who is staying for the weekend with Lisa. It is a shock to the system waking at 4.30am, but the rewards will be worth it. We start the day by filling flasks and loading the rucksack with home-made muffins. When we step outside, it is dark and silent with a heavy, overcast sky but thankfully dry.

We take the car through Cheddar Gorge, wending our way ever upwards to a point where we park at the wood adjacent to Black Rock. Having put on our boots and warm clothing, we wait by the car for the first light and indeed the first sounds. After a pause of five minutes, we hear from the north a female tawny owl which is followed by the call of a pair from the east and another pair from the west. The lone female is soon joined by her mate, and the parliament of owls call incessantly to each other before settling down for the day.

The sky to the east is beginning to lift its head from the pillow and is heralded by the melancholic call of a robin; always the first to rise. Now the chorus starts, and the song thrushes and blackbirds give voice to the occasion, with each trying to outcompete its neighbour and potential rival. The blackbirds' fluty call trickles through and across the woodland like a meandering river whilst the song thrushes perch on high, repeating their verse three times over just in case we miss it the first time. These early soundings have us enraptured, and the timbre of the woodland dawn is all around, with each individual becoming harder to identify. Just as I think it impossible for a single bird to hold a tune with the power to rise above the throng, so the distinctive trill of a wren lifts clear of the mass and continues to do so.

We agree it is time to move into the wood and start our walk. Just as we reach the gate, a pair of blackbirds become highly agitated and pitch forth with their alarm call. We speculate as to what the problem might be, thinking it might be an owl; however, it becomes clear when we hear the call of a jay, that most colourful of woodland corvids and known raider of nests to feed its own growing brood.

In the wood, the light is subdued, but there is enough for us to see our way, all the while looking for roe deer and still accompanied by the society of choristers. We exit the wood onto Black Rock which is being surveyed for loose rocks and made safe. As we turn the corner, we come upon the Exmoor ponies brought here to help protect the landscape and keep scrub to a manageable level. They are not tall ponies, standing at twelve hands, with a brown coat and distinctive mealy markings around the muzzle and eyes. I am particularly fond of the Exmoors as they are not only an ancient breed but also local. They seem unconcerned by us, providing we keep our distance, so we move on, still heading north-west, leaving Velvet Bottom to our right with rabbits scattering left and right.

Eventually, we come to Long Wood, its entrance marked by a lone song thrush calling from a tall ash tree and amplifying its value by casting his voice onto the surrounding valley. Entering the wood, we notice fresh deer slots, indicating their presence, and so we climb the path that takes you on a trail deep into the wood which is fresh with the lime-green foliage of newly emerging leaf. The carpeting of bluebells is a good seventy per cent in bloom and offer up a beautiful combination of colour and scent. There are small dashes of white where the wild garlic is about to flower.

As we climb the muddy path, we hear the distinctive laugh of a yaffle, more commonly known as a green woodpecker. Once we reach the top of the wood, we halt by a drystone wall blanketed with aged mosses that are damp and soft like a newly soaked sponge. Peering over the wall onto the adjacent meadow, it is hard to see any distance due to the thick, dank mist, the quality of the light being like quicksilver as it shifts in the fresher wind and envelopes the hedged trees. Further down the line of the wall, two cock pheasants dash forth, taking their dispute into the field and, when noticing us, slowly stalk back, brothers in arms hoping to go unnoticed.

We re-enter the wood and carry on along the path, not sure where it will take us, when Sam notices a different print in the mud. From the broad pad, toes and distinctive claws, it is clear it has been made by a badger. Just a little further along, there is an animal track leading down into the thicket, and we no longer see evidence of the badger on our thoroughfare.

Eventually, we retrace our steps and gingerly leave the wood, trying not to slip on the muddy slope, and return to the lone song thrush who is still calling from his tree. We explore onto the higher slopes of the Mendips where we take our breakfast of muffin and tea and eventually head back down to the Long Wood entrance. As we take one last look at the wood, we hear the loud drumming of a great spotted woodpecker sounding out his territory. An old trick I have used before is to return the call by using stone on wood and drawing the bird into the open so, without luck, I try it and inadvertently flush a roe deer from her cover. The deer is in excellent condition as she strides purposefully through the wood using the trees as cover, and it is clear she is heavy with a fawn.

Heading back to the car, we notice the ponies high up on the ridge and silhouetted against the leaden sky, the knarred and twisted hawthorns stunted by the conditions, adding a timeless quality to the scene. Passing through the first wood some three hours after we started out, there is still some song, but it is far more subdued as the birds go about their daily chores.

If there is one thing more satisfying than experiencing a dawn chorus on your own, it is the sharing of it with loved ones. The magic and tranquillity of the moment never leaves you, and to see Sam become absorbed in all the detail that such an unsullied walk offers makes such a moment extra special.

Draycott Sleight

It is the last weekend of July, a hot summer's day and the end of an equally hot week. Lesley and I decide to make the most of it and go for an afternoon walk; we choose Draycott Sleight as our destination: a south-west facing reserve that has a steep slope leading onto the more level ground of the Mendips, owned, and managed, by Somerset Wildlife Trust. This is a lovely reserve as, during the autumn and winter months, it holds a beauty in its bleak and blustery demeanour and yet, during July and August, there is a sensitivity with its rich collection of butterflies – some of which are uncommon – as well as the various wildflowers and grasses. The name Draycott Sleight is an interesting one as the origins are from Britain's Norse and Old English past. 'Dray' refers to shelter for the pulling horses and 'cott' (or 'coet') references the woods; therefore, 'Draycott' is the shelter in the woods. 'Sleight' is an Old Norse word ('sletta') meaning level field or meadow, which would presumably refer to the Mendips.

Draycott Sleight sits between two areas of national importance for wildlife habitat – the Somerset Levels and the Mendips – and it is this midpoint we are in today that link the two. Walk the slope to the east, and you are in a landscape that appears minimalist and yet is complex in the way it supports its dependants. Walk far enough towards the south-west, and you are on the Levels with its lush pastures and criss-crossing rhynes.

The moment we pass through the gate and look up the slope, we see a flurry of activity from the chalkhill blues, a butterfly that is rare in this region, given that it relies on free-draining limestone and chalky soil that supports its only plant source: horseshoe vetch, a small ground-hugging plant with delicate yellow pea-like flowers. As well as the horseshoe vetch, there are tall, wispy stands of dried grasses that host meadow brown, marbled white and small skipper butterflies.

Climbing upwards and being careful to only tread on the rocks or the carefully crafted sheep track, Lesley and I make our way in amongst the action. I focus on one male chalkhill blue that has set up his territory around a self-heal plant that has a short, spiky stand-in flower. The proud owner of this territory perches on his elevated position and waits for a lovely lady to pass his way or to see off a wandering rival, something accomplished at speed. Depending on how this alert animal presents himself determines how much of his blue coat is displayed, and if he gets the angle right, then the depth and richness of the colour is captivating. In this small patch of protected land, there are several powder-blue dandies defending their own patch with the occasional flurry of activity as one crosses into another's domain.

Time to move on, and so we climb the escarpment, heading for the higher ground of the Mendips where the gliders are launched towards the nearest thermal. As soon as we come off the protection of the sleight and reach the upper levels of the Mendips, it is a more noticeable rugged terrain with a brisk south-westerly wind to accompany it, which in turn is ridden by kestrel and buzzard alike; both scouring the land for a ready meal. The views from up here are stunning where the clear air allows you to look out across the Somerset Levels and beyond the Bristol Channel into Wales.

The wildflowers surrounding us are wonderful, free of artificial fertilisers and any form of intensive farm management and left to flourish. There is horseshoe vetch, pink clover, harebell, goat's beard, devil's-bit scabious, bird's-foot trefoil and wild thyme in abundance. Picking a bit of wild thyme and rubbing it between your fingers gives off a deep aroma fit for a winter stew.

We skirt the edge of the field and watch the gliders winched to the heights and, as we return in a wide arc, we find our way back onto the path that eventually leads to the upended limestone sleight. As we walk the path, we come across a small, deep pond patrolled by a four-spotted chaser: a fearsome-looking dragonfly that zips about its territory with menacing intent.

On we go, and now we pass a tall stand of one hundred beech trees. These trees were planted as a hedge, but that must have been well over a hundred years ago as their girth, now wide and well fed, is beyond restraining. It is clear to see that it was a once a hedge as they sit on a weathered bank, and there is evidence they were layered and coppiced as they have, in their rebellion, thrown up many trunks from the thick, low-lying bases.

We are back to where we started and have hardly seen a soul. I love a landscape that talks back to you, and this one quietly calls out to me. Who planted that hedge all those years ago, and how must the land have looked to them? What forces of nature upended the carboniferous limestone that now hosts a beautiful butterfly? And how many people past and present have found solitude in this peaceful haven?

Glen Licht

It is the beginning of October, and having already been in the Highlands of Scotland over the last few days, Lesley and I decide to go on one of our favourite walks. We are walking Glen Licht, which is a three-and-half-mile walk that follows the River Croe. Glen Licht forms part of the Kintail range, and our walk starts at the ranger's office and finishes at Kintail House which is a small lodge used by the University of Edinburgh. The weather is not good, with a lot of heavy rain interrupted by the occasional burst of sunshine, but I do not care as I am fully waterproofed, and it feels exhilarating to be a part of raw nature and not just a hostage to it.

We leave the car park next to the ranger's office and continue up the road, passing some beautiful Scottish Blackface and Cheviot rams grazing in a paddock and looking fit and primed to do their duty. Passing through the gate next to the activity centre, we enter the glen and before us is a breathtaking view of the Five Sisters and Meall an Fhuarain Mhoir mountain ranges with the River Croe to our left. We follow the river for several hundred yards to a point where it is cascading in full spate over boulders and falls in spectacular, noisy fashion, leaving me feeling disorientated as the water looks so solid and the rocks so vulnerable. The last time we were at this spot, we watched a dipper diving for caddisfly larvae; to do so today would be suicide for the intrepid bird.

Moving up the valley, the river is full bodied yet calmer, and we enter the head of the glen, which is a great example of glacial action, the steep-sided mountains leading to a valley floor which in turn leads onto Loch Alsh via Loch Duich. The riverbed, mountains and footpaths littered with an assortment of stones, rocks and boulders which have probably been transported from miles away.

The ground is very wet, and in places the mud deep and slippery, watered by the burns running off the Five Sisters, some of which are broad with mountain-clear water. The mountains tower above us, shrouded in mist, and the magnificent red deer stags call from all around, augmenting the atmosphere. Occasionally, the clouds clear, and great shafts of sunlight travel westwards down the glen and, like enormous spotlights, they scan the face of Meall an Fhuarain Mhoir, and it is then that the tapestry of colours stand out. There are the remnants of the purple heather, ochre, vermillion, crimson and various shades of browns and copper, blending on the most awe-inspiring canvas and brought to life by the sun.

Throughout the walk, there is one stag that is constantly calling from the summit of Meall an Fhuarain Mhoir and yet I cannot see him due to the thick mist shrouding the top. The rut is in full swing – there are stags seeking a harem – and it is clear this animal is staking a claim. Approximately two miles into the walk, the mist thins, and I look for the persistent and elusive stag. First I see a hind and then the stag, head back and bellowing, when slowly, unsummoned, a golden eagle drifts into view through parting mist and rides the updraft. This is nature at its absolute best: imperious mountain briefly in communion with stag and eagle before each slip back into anonymity.

Having ambled up the glen, we eventually arrive at the lodge, and the sun has now been out for a sustained period, which in turn has produced delicate insects drifting over the water, encouraging the fish to leap from the calmer pools.

It is possible to continue a circular walk around Meall an Fhuarain Mhoir; however, this a longer and more testing walk. We are content to return along the same path. With the light still good, we set off and see a further two stags on the Five Sisters as well as another golden eagle. The weather constantly changes, and we are occasionally hit with squally showers which clear to sunbursts of light.

We have walked this glen across all seasons, and it never fails to captivate me; it is an absolute favourite. The mountains with their changing moods, varying light and colour, combined with the wildlife, leave me feeling humbled and completely at peace. When I am physically no longer able to visit this magical place, it will always reside clearly in my memory.

Garston Wood

An afternoon in May and I am midway through spending one year visiting as many woods as possible, and Garston Wood in Dorset, which is owned by the RSPB, seems to me a suitable candidate to include.

I arrive at the wood at 5pm as I want to see plenty of wildlife before the sun sets. On arrival, there are a group of scouts who, in their excitement, are milling around making a lot of noise. The scout leader apologises; no need as it is great to see so many youngsters showing such enthusiasm for the wild – we need more engagement of this kind if we are to protect our natural world for the future.

The very first thing that strikes me as I enter the wood is the mass of bluebells which are late flowering. I have never seen such a thick carpet of iridescent blue which lights up in places where the full sun breaks through, and the heady scent of the flowers is equally intoxicating. Contrasting the bluebells is the wild garlic which competes for volume of flower and scent. As the sun drops lower, long shadows develop, and accompanying pools of soft light settle on the blue and white carpet.

This is a wood that comprises of oak, maple and hazel that was traditionally coppiced, and to encourage the continuing existence of the wildlife, the RSPB and volunteers continue the ancient tradition. The newly opened leaf cover is young, transparent and uncurled, to provide a summer of life-giving nourishment to a multitude of insects, birds and mammals. This is a beautiful, soft-edged wood that cradles life and is not at all oppressive; I feel completely at peace here.

Settled in among the dominant blues and whites are the more delicate lesser celandine, wood anemone, yellow archangel, bugle, common spotted orchid and wood sorrel. These elusive and pretty sideshows add depth to the overall image and encourage your eye to wander and are a strong indicator that this wood is not just beautiful but also aged.

The evening chorus is in full flow, and the bird that dominates is, as usual, the song thrush, and he is matched by competing neighbours with the same repertoire. It is great to hear so many song thrushes as this is a bird in decline, and our countryside would be a sorry place if numbers dropped any lower. The other bird that stands out is the blackcap with his jaunty song catapulted into the wood at full volume, and if you should be so lucky to see the bird, you will notice he wears his black beret with the same carefree abandon.

The wood is renowned for its marsh tits, and it is not long before I see one which is a first for me, a beautiful bird that keeps low as it flits from one tree to the next. A little further on, I hear a garden warbler and, after much searching, catch sight of it; again, this proves to be a first for me, and to add to the beautiful melody of this vocalist, the occasional willow warbler joins in.

Another great bird to mention is the cuckoo. I frequently hear the cuckoo on the watery Somerset Levels where they shadow reed warblers, but as a child, I was more used to hearing them in the depth of the countryside, across the wooded coombes of Devon. It has been decades since I have heard a cuckoo in the countryside, so to hear one this evening transports me back to my childhood; peeling church bells would have completed the memory.

The sun is now dropping closer to the horizon, and the resident fallow deer begin to stir as they hunger for the open pasture. Travelling down one of the woodland glades, I see three cross in front of me and stand on a low bank with one silhouetted between two trees. Others to the left catch sight of me and are hesitant; in the end, they decide to turn back. I step onto the same bank to see where they are heading, and there are a dozen stood in a field of grass looking back, wondering where their companions have retreated. I decide to investigate and go in search for the ones that turned back, eventually finding them in the protection of the wood, knee deep in garlic and with the shafts of mellow light picking them out. Suddenly, a dog appears next to me and the deer spook and make a dash for the field, leaping the sunlit path like gazelles on the African plain. These noble animals of the wood belong here, and I promise to return in the autumn to hear the bellowing of the bucks.

As I amble back to the car, I am in for one more special

treat. It is well known that one of the animals that I am attached to is the brown hare, and whenever I see one, it truly cheers me on. So, I am walking the path in a westerly direction with the setting sun in front when two hares nonchalantly saunter out in front of me, no more than twenty metres ahead, and proceed slowly along the same path. Normally with hares, you do not move a muscle, as their exceptional vision will spot you and they will bolt; however, on this occasion, I gingerly decide to follow and still they take no notice; it is as if they know I will do them no harm. Eventually, they turn left out of the wood into a field of what looks like broad beans. What a privilege to be a part of the wood where you no longer feel like an intruder but as an accepted component.

This beautiful wood with its rolling glades, soft carpet of wildflowers and stands of elderly trees has a wonderful, inclusive feel, and the birds, deer and hares complete the sense of wholeness. I have visited many woods recently, and this is one of my favourites.

Tranquil moor

Glen Licht

Triple Buttress of Coire Mhic Fhearchair

Cornish coast looking to the Rumps

Callanish Standing Stones

Aurora Borealis seen from the Isle of Lewis

Sunrise from Roughtor

Webber's Post

It is the beginning of June, and this afternoon I am off on one of my woodland adventures and once again decide to visit somewhere new. This afternoon I am visiting Webber's Post on Exmoor which is an old wood that has settled into a beautiful coombe, home to ancient oaks. Webber's Post is part of the Holnicote Estate and is one of the largest wooded national nature reserves in England and one of the UK's temperate rainforests, sometimes known as Atlantic rainforests, special for their climate, flora and fauna. The small river that ambles along the seam of the coombe is called West Water and runs clean and clear.

From the car park there are magnificent views of the whole wooded estate that drifts over hills and descends into deep coombes, and towering over this verdant scene are the imposing moors. This is the very essence of Exmoor and captures why I enjoy visiting it so much in preference to moorland cousins to the south.

I start by walking the road towards Cloutsham and the first to greet me is a bold willow warbler brazenly sitting on a tree, sharing its delicate song. At this uppermost level, the dominant tree is the birch, standing over a rich carpet of bilberry. Other plants I pick out on the walk include germander speedwell, common dog violet, lesser spearwort, common cow-wheat, creeping jenny, and bursting through them all are large crowns of fern, uncurling and bejewelled.

The light is fantastic, and the wood responds with a benevolent reflection that creates a pallet of colours that change density with the oscillating light and shade. It takes time to adjust to a new environment, and it is something that cannot be rushed; you must give the place time and allow yourself to be drawn in. Slowly, I descend, keeping all my senses open to shifts and changes and, gradually, I begin to see and hear things, some familiar, others not. One of the earliest birdsongs is one that I do not recognise, and after much searching, I see a pied flycatcher, another first! This is a small bird, similar in size to a chaffinch, and one that is full of character and looks impressive in his dinner suit.

Carrying on down the hill, I hear the familiar clatter of a great spotted woodpecker and spy it climbing along the branch of a mature oak that has had its top savagely torn away. Descending the trunk of the tree, there are four holes of equidistance drilled for nesting: four years of breeding, which is evident when I get closer and hear the noisy woodpecker chicks begging from within one of the holes. The dead standing tree is in a sorry condition but supports a plethora of life long after its own demise.

Further down and the wood is dominated by oak with impressively aged trees that have hosted many and stand as gatekeepers, reminding man and time that change is not always a good thing. Finally, I reach West Water, which nestles at the bottom of the coombe and slides uninterrupted over the road towards the meandering river on the other side. Where the water crosses the road, I gingerly cross the slippery stones and hear the call of a yaffle, that could easily be laughing at this stranger walking like it is his first step. Having walked on for a few hundred metres, I decide to turn back as I want to walk alongside West Water and look for dippers and trout. On my return to the river, I spot my second pied flycatcher bringing insects back to his nest: a small hole in a sturdy oak. I am treated to some great views as he brings meal after meal of insects to his brood. I watch this wonderful bird for ages, and as I do so, six Exmoor ponies appear like spirits travelling on an unseen, elevated path to the back of the wood, adding to the ethereal mystery of their coming. I am very fond of these ponies as they are an important part of our local history and have a deep, integrated connection to Exmoor.

Tearing myself away, I return to West Water and follow its winding path, and as I cross it several times, so I am led into the depths of its secrets. The mature trees, dappled light and gentle river demand that I travel slow, respect my surroundings and miss nothing. I do not see dippers, but brown trout scatter as my reflection is cast over the lively water.

Eventually, I reach the point where I must leave the intimacy of the coombe and climb the steep escarpment back to Webber's Post. It is a tough climb, and I think my heart is going to leave my chest!

I love these Exmoor coombes with their aged denizens standing tall, proud and bearded with dappled light drifting through their outstretched limbs. The birdsong is at its finest point in the year, and the migrants returned from their long journey add their own harmonies to the more familiar vocalists. The clear water meanders into the distance, bringing relief and nourishment to the surroundings, and my need for a tranquil afternoon on Exmoor is met.

Andalucía

Lesley and I are in Andalucía, close to Ronda, for two weeks, and today we are going on a guided tour of the local area to do some birdwatching. I have been looking forward to this day for months as it is that long since I booked it. We are joining Peter who not only lives locally but has spent his working life in the world of natural science and education.

The morning is bright and sunny with not a cloud in sight and not a hint of a breeze. It is quickly apparent that Peter's knowledge is second to none, and he is very easy company. Our first mini tour are the lanes surrounding our accommodation to which Lesley and I subsequently return to, a real treat to have this on our holiday 'door step'. Peter is quickly picking out birds, and one of the first is a linnet, which is a gorgeous finch with a rose-tinted breast and one that I have not seen in over twenty-five years. Onwards and we see serin, woodchat shrike, crested lark, thekla lark, short-toed eagle, a beautiful red rump swallow and various warblers, including the melodious warbler.

A special bird that is in abundance is the dashing and colourful European bee-eater, one of those birds on my must-see list established when I was that nine-year-old boy. The bee-eater is striking in so many ways with its multicoloured feathers of blues, yellow and rusty reds, the eye banded and framing a piercing, red-eyed stare. Taking to a perch with a commanding view, it will confidently rush forward to chase down and catch insect prey that others would avoid, including bees and dragonflies. We return multiple times to watch these beautiful, charismatic birds.

Before leaving this, our local patch, we come across another gorgeous bird, the turtle dove. This small dove is not only a lovely bird to look at with its soft coloured plumage, flecked wings and neck band of silver and black, but it also has the most soothing of summer calls that transports you back in time; what a tragedy that in the UK it is close to extinction!

Travelling towards Setenil, we take a detour and turn left, driving up a rutted lane towards an escarpment that has wonderful panoramic views of Montecorto, Grazalema and Lake Zahara. Rising from the heated ground far below are columns of invisible thermals, ridden by booted eagles and griffon vultures as they effortlessly conserve energy drifting upwards and over us. This is so captivating as these majestic birds come so close on the rising breeze that you could be there with them.

The tour continues onto Dolmen del Chopo and craggy mountains near a disused dam. En route we see a short-toed eagle, rock sparrows and a beautiful blue rock thrush.

We enter a village where a pair of Bonelli's eagles have their eyrie but sadly this year no chicks and we did not see the eagles. However, not to despair as the terrain is lunar in appearance and therefore promises to offer up different and unusual birds, and I am not disappointed as we see dapper black wheatear in the rocky landscape. Further on and we turn a corner into a valley that is wide, far reaching and fertile and nestled within the overheated mountains; these are the Líbar plains, and they are breathtaking. The contradiction in scenery is stunning. The birds are equally wonderful, including choughs, red legs and beaks adorning shaggy black coats, tucking in their wings as they stoop and glide, all the while calling playfully to each other. Spotted flycatcher dropping down from a favourite perch, a balancing flick as they snatch unsuspecting prey, before gliding back to the same perch. Cirl bunting, corn bunting, lesser kestrel and black-eared wheatear all playing their part in making this a day to remember.

We decide to have lunch in Benaoján and, travelling through the village, we see a male golden oriole, a dandy, garlanded in yellow and black with a finishing flourish of red beak – surely this birding expedition cannot get any better! I first heard a golden oriole many years ago when on holiday in Jazeneuil, a small village near Poitiers, France, surrounded by a river-channelled woodland populated with oak and shimmering poplar. I would get up at dawn and go out to capture in watercolour the village church, with my chattering eldest son Sam as company, and rising from the wood would float the soft, fluty utterance of the golden oriole. A smooth-edged and memorable moment: dawn, birdsong, paints and talkative son. The lunch incidentally was a fantastic pork stew in a restaurant full of locals.

On our way home, we stop twice along the Río Guadiaro to see little ringed plovers and, further along, some olivaceous warblers, crag martins and alpine swift. The final stop is to see a hoopoe nest site, to which I return to take pictures.

Throughout the day, the one thing that stands out alongside the birds and breathtaking scenery are the wildflowers. These are not only on the margins but also in the crops, something not seen or tolerated in the UK. The flowers include fennel, wild carrot, peony, love-in-a-mist, poppies, various thistle and wild delphinium. To see such abundance of wildflowers and the wildlife that thrives is heart-warming.

It has been a wonderful and enriching day, and the whole experience reminds me of my childhood and the joy of first discovering birds in a natural and unspoilt environment.

Skomer

A Father's Day treat in store with an opportunity to see the great seabird colonies on Skomer Island off the Pembrokeshire coast. As it is the 15th of June and the colony at maximum capacity with the prospect of good weather, the opportunity for some great sights is looking equally promising. We need to arrive for an early start which means leaving Cheddar at 5am to arrive at 8am to book a slot on the 10am boat.

Adding to the treat, Hallie, our eldest granddaughter, is joining Lesley and me for our day out. Hallie shows a lot of interest in wildlife and not only is it enjoyable having her with us, but we also get an opportunity to encourage her early interest.

Having eaten our jam sandwich breakfast on the drive, we have an easy journey and arrive at the car park for 8.15am. We are fortunate in that we get one of the last tickets for the 10am crossing, which means a full day on the island. We wait two hours to reach the island which is bearable for two grown-ups but less so when you are five. Next to the slip road leading to the jetty there is a basic museum with a few exhibits, but what is most fascinating are the two pairs of swallows diving in and out of the door to feed their young. Hallie is fascinated watching these beautiful hirundines dip under the door, arc up to their nests and occasionally take a 360-degree circuit around the room, always accompanying the visit with a barrage of chatter.

Finally, we get on the boat, all fifty of us squeezed on like sardines in a tin. As we pull away from the jetty, a peregrine contemptuous of us lazily wings her way down the coast using deep laconic wing beats that belie her strength, destined for her eyrie on the island. Once out onto open water, the sea gets a little choppy but nothing heavy or concerning, and as we approach the island, there are auks rafting everywhere, great gatherings of puffin, guillemot and razorbill bobbing across the surface. When the boat gets too close for their comfort, the birds dive for safety where they can, taking to the depths with ease. The sea is alive with birds and is a wonderful and enthralling spectacle, and the thriving numbers on the surface are indicative of what wealth of life and nourishment lies below and out of sight.

We eventually land and climb the steep steps to the Wildlife Trust team who give a briefing ensuring we are aware of the island's sensitivities and for us to respect them. Briefing over, we decide to head for the old farmhouse in the middle of the island where we stop for a short refreshment before heading off along one of the tracks that lead to the coast. The island is awash with great seas of red campion roiling across the landscape, a real treat for me as I love to see this, one of my favourite flowers. Sat in amongst this pink ocean is a lesser black back gull that might look out of place, but imagine substituting the pinkness for one of oceanic blue, and it looks perfectly at home.

One of the more disturbing elements of our day out are the number of dead Manx shearwater on the paths and in the scrub, contorted bodies that have met a violent end. Manx shearwaters spend their days out on the ocean fishing for their hungry youngster and only return under the cover of darkness when they can dash to the safety of their burrow, reducing the risk of being ambushed by gulls and, in particular, great black-backed gulls. Remarkably, there are estimated to be 316,000 pairs of Manx shearwater on Skomer (2011 whole island census) with another forty thousand on neighbouring Skokholm, which together constitute over half the world's breeding population. Predation by great black-backed gulls would therefore seem insignificant, but I do not think the starving chick in the burrow would quite see it that way! Walking around the island, it is clear to see the gulls hang around in small gangs, self-assured and muscular, waiting for the right moment to cause a little trouble.

Passing the gulls and heading towards the coast, we pass great flocks of linnets, beautiful birds with soft, rose-tinted feathers and a tinkling call. To contrast the lightness and delicacy of this beautiful finch, I am treated to the mournful call of a curlew which is one of my favourite sounds and certainly belongs here on Skomer.

On the rugged coastal fringe, there are thousands of

seabirds, in particular kittiwakes, razorbills, guillemots and puffins. The portly puffin is the most striking bird with its colourful beak and black-and-white dinner suit for plumage. The puffins that are not at sea or down a burrow hang around on the terraces like elderly gentlemen considering the trials of their day; the only thing missing is a pint and smouldering pipe.

One piece of bird behaviour I did not expect to witness is a buzzard hunting out on the cliffs. I am accustomed to seeing buzzards wheeling high over woods, copse and verdant pasture; however, this bird is out over the cliff and hovering kestrel-like, and it is clear to see that it is in a hunting mood. It is a fascinating piece of behaviour, and it is clear the buzzard has adapted its hunting skills to take advantage of the abundant young birds on the seabird ledges.

Along the way, we will occasionally stop to look at the birds, and Hallie joined in every time, lifting her binoculars to survey the scene. I believe she is interested in the birds and is serious in looking for them through her glasses.

When we stop for lunch, we have a clear view of a cove which has a settled, glassy sheen, allowing us the opportunity to survey for cetaceans and, in particular, porpoise, which are known to hug close to the island. The best way to find porpoise is to look for overhead gannets who use the mammals as beacons, leading them to a food source and, sure enough, I find half a dozen gannets circling quite close to shore, and below them we spot a porpoise with calf. Lesley and I always get excited when we spot cetaceans, and this sighting adds to a collection of wonderful experiences.

Having refuelled, we make our way to the Wick, a wedge-shaped cove on the island with a deep-set cliff face, and it is here that we see the greatest abundance of puffin. It is amazing to be standing on the clifftop watching the birds skimming in on tiny wings, beating furiously to keep them aloft. Once they land, the puffins make a dash for their burrows, ensuring the beak full of sand eels reach the chicks and are not stolen by menacing pirates.

Leaving the Wick, we head back towards the landing site and on the way pass High Cliff, a cove where there are a large colony of guillemots and razorbills. It is the guillemots that catch my eye as there are many birds on the cliffs which include larger chicks. Directly below the cliff there are large rafts of birds that include male guillemots calling to their chicks high above them. It is more than I could have hoped that I might see one of nature's truly wonderful wildlife spectacles: chicks joining their fathers on the water. Having seen the spectacle on television, I never hoped to see it for myself, but I am not disappointed as two chicks launch themselves off the cliff and clumsily glide towards the sea like a child taking its first steps. These flightless chicks will now be escorted by the male on a long journey to open sea off the coast of Spain and France where they will overwinter, learning the skill of fishing in deep water.

We arrive at the landing platform exhausted but thoroughly contented. Hallie has been good throughout the day and has shown a genuine interest in the wildlife. We eventually board the boat which has reduced numbers due to the sea being choppier than when we arrived. Once out into the channel between the mainland and the island, it is clear the sea is quite rough. Lesley and I are sat in one corner with Hallie wedged between us where one minute the gunnels are low to the water and on the next, we are several feet in the air; it is like a fairground ride and Hallie is loving it.

When we finally land and make our way back up the hill, I am presented with one last treat: the sight of a chough. I had heard the birds at one point through the walk but could not see them. Choughs are real characters, and they have this distinctive sociable call that sounds like *chack*, and as this bird flies past, it calls out as if to wish us well on our journey home.

I have had a wonderful day that met all my expectations and was made even more enjoyable knowing that Lesley and Hallie had a great time.

Ham Wall

I had recently visited Ham Wall on the Somerset Levels to hopefully see tawny owl owlets in a box overlooking the Avalon Hide. I did not see the owlets but was not disappointed by a pair of barn owls coming out in the early evening and with good views.

Discussing it with Lesley, we agreed to revisit and take Hallie with us. She has always wanted to see an owl, and her interest has grown over the years to the point where she can recognise many birds. It is a month before her eighth birthday, so she is the right age for the interest to develop.

It is a fine weather forecast on the 19th of May, 2017, so we agree to pick up Hallie, take her to the Somerset Levels and for her to stay with us afterwards. We head to Ham Wall under what looks like an ominous sky with brighter prospect from the prevailing west. We arrive at 7pm with plenty of layers, as sitting in a hide for hours can be a cold experience. The walk from the car to the hide takes approximately twenty minutes, and we are stalked by thousands of midges, fortunately the non-biting kind! Hallie is extremely excited and full of bounce and chatter.

Just as we get to the hide, there is a massive downpour which not only wets the surrounding area but severely dampens my spirits as I was really hoping for this special birdwatching experience to go well for Hallie, with it being her first outside of the back garden. Owls will not come out in the rain as they have poor waterproofing, and they cannot hear their prey over the noise of rain, meaning they are unable to hunt using their acute hearing. The rain passes as quickly as it arrives, and the promise of clear weather from the west does return.

Inside the hide, there are half a dozen birders with space for more. We settle into our seats overlooking the reed bed where we see lovely sights, including a family of swans with young cygnets and great crested grebe with the stripy young riding piggyback. It is lovely to see Hallie at an open window taking in the sights and sounds of this wild and enchanting environment. Within the hide, there is evidence of a barn owl roosting overnight high up on a beam with its droppings whitewashed below.

At approximately 7.45pm, we look, with anticipation along the rhyne towards the wood on the east side of the hide as this is where owls, tawny and barn, have previously been seen. And there it is, a barn owl, and without the use of binoculars, the first to see it is Hallie, an exciting moment for us all but especially for her. The owl skirts the wood and, as it slips out of view, we all rush downstairs for an unobscured look along the path to where it disappeared.

The owl quarters back and alights on a rustic bench in that classic pose of patiently looking out across the reed bed.

Others in the hide decide to take a chance and walk to the wood in the hope that they will get better views of the owl, leaving Lesley, Hallie and me alone in the hide. We decide to stay in the hide and take the gamble the owl will reappear. After a long wait, during which I begin to doubt I have taken the right decision, we are rewarded as the owl again skirts the wood from right to left, only this time it is clearly carrying a vole, limp and bulky, back to the nest to feed a growing family of owlets. We all have great views with our binoculars, and Hallie really understands how to focus on the bird and prey.

As well as the barn owl, we also see the male and female marsh harrier that drift and float kite-like across the reed beds on outstretched wings, and I am lucky enough to see a merlin as it appears from nowhere and, wasting no time, dashes low over the hide.

We finally leave the hide at 9.15pm contented and happy. The midges are there to greet us and escort us off the reserve, and we slowly make our way home to a glass of warm, sugared milk.

It is so important to introduce the younger generation to the joys of the natural world as soon as they can comprehend it. It needs to be fun and magical, and they must be allowed the freedom to explore and wonder at the brilliance of it. Our precious blue marble of life needs friends, many friends who can help protect and treasure it and that must come from each successive generation. Lesley and I get as much pleasure from sharing the owl experience with Hallie as she does herself in seeing it.

Glen Arnisdale

It is early June and another holiday in the Scottish Highlands. This is home from home for me and Lesley and again staying in Glenelg.

Today we are going to try a different walk along Glen Arnisdale, finishing at Dubh Lochain. When visiting last year, we went part of the way on this walk and today intend to go much further, so with fair weather promised, we plan to complete the return journey of just four miles. We park the car at Coran and pass through the farm gate adjacent to the River Arnisdale which at this point is shallow and comfortable as it spreads itself across a bed of rocks and pebbles. The grass is long and lush and the ground slightly soft underfoot.

As we crest a small incline, we see in the distance a gathering of stags; their coats look moth-eaten and their antlers covered in the blood-engorged velvet that supplies the new bone with the nutrients to help them quickly grow strong. It always amazes me that new bone can grow so quickly, for it to intimidate an opponent during the autumn rut and, if necessary, to spar with. Seven months later, all that energy-sapping bone production is discarded for the process to start again. It is not long before Lesley and I are spotted, and where we thought there were a few stags, increasingly more appear until, eventually, twenty stand wary of us.

The stags are near a bend in the river at a point where the wide, expansive bank is shingled and grey, and the water carelessly meanders by. The stags are moving from right to left away from a grove of alder along the riverside towards a pasture where they can more easily manage the threat of our presence. To make the pasture they must leap a fence which they make short work of. I love watching deer at any time of the year, and this bachelor group do not disappoint. We move on and leave them in peace, only to discover there are four that have become separated, and they in turn take to the opposite direction where they eventually stand observing us from afar in boggy ground set with spangling cotton grass. In time, these four, along with the original group, join a herd on the mountainside, creating one large group of approximately forty stags of different ages and sizes, all keeping us at a safe distance.

The walk is easy and flat and, having gone a little inland, we again join the river which snakes its way westwards and occasionally tests itself over pebble-rich terrain; for company, grey wagtails dance and flick, their tails as if keeping beat to the humble tune of the river as they reach for waterside insects. We pass a bank to our right that has old and new sand martin burrows and, unexpectedly, a diligent parent skitters from its nest entrance and dashes to the river for more winged supplies. What hope for the multitude of invertebrates with these two denizens of the river dashing and hawking for a meal.

Entering a lush part of the walk with beautiful trees to our left and to our right open terrain leading up to mountaintops, we take our time to enjoy the scenery and the wildlife. One large piece of land on the open side is fenced off with tall deer wire preventing their access, and in this area, the trees and plants have been left to flourish, and what a difference. There are several species of native trees; the heather grows uninhibited; and there are no ugly plantations of spruce. I love seeing the deer, but this fenced-off experiment goes to show there are too many herbivores overgrazing the hills, and that includes the abundance of sheep. The raw and naked beauty of the mountains should not be lost to some rewilding ideal, but equally, there must be a balance to allow native flora to establish itself, which under the current circumstances is not happening.

Before slowly descending into a small, shaded wood that sits close to the river, we meet a sprightly, elderly couple coming from the opposite direction; we exchange greetings, and my thoughts wander to hoping I can enjoy this wonderful country at their age. Incidentally, these are the only two people we meet on the whole walk. At the entrance to this next wooded section, there is the awful smell of decaying flesh produced by a dead deer behind a fallen log with just a few bones and its hide remaining. Moving upwind, the light in the wood is beautiful, and the trees zing with dappled tinges of green; drifting in amongst the shifting luminosity are the descending notes of competing willow warblers. There is a scattering of pretty flowers along this walk including pignut, germander speedwell and dog rose.

The walk levels out for a brief period, before slowly ascending through oak, birch and alder until we reach a large step of scree lunging skywards. Nothing for it but to pace ourselves; we slowly scramble upwards, and just as you think you have reached the top, so it bends sharply to an angle and continues upwards. This challenge is repeated until, finally, the top is reached, breathless, but the effort is worthwhile as the views down the glen to Loch Hourn and beyond are wonderful.

The walk is now considerably easier, and it is along here that we see new and interesting flowers, ones I have never seen before such as common lousewort and herb Robert. There are so many flowers that, with my limited knowledge, I can only pick out a few; however, there is one that stands out, and which I have never seen before: the lesser butterfly orchid, a beautiful spike of white flowers that stands out from the crowd.

A steady stretch and our destination is in sight, a waterfall and path leading to the loch. Having escaped the loch, Dubh Lochain, the clear, fresh water, with force, cascades over a lip of boulders and what appears to be a broken stretch of concrete dam and, upon reaching the base of the fall, spreads its mass over a broad stretch of smaller rocks and eases itself towards the beckoning Loch Hourn.

Away from the sheltered valley and bowered wood, we reach Dubh Lochain, where there is quite a breeze pushing across the water, and the towering hills bordering the water offer little resistance. One of the great treats when walking in Scotland at this time of year is the number of cuckoos you hear and, as we now take a rest, we hear our third, and I take the opportunity to record it calling. Having sheltered, taken in the view and eaten our lunch, we just quietly contemplate having this beautiful place to ourselves.

We eventually stir and take the same route home. Along the way, there are deep gouges in the path filled with fresh Scottish rain, and in one of them, I spot a newt and tadpole; it is incredible how nature carves out a home in the most unremarkable spaces.

The walk home is lovely and the sense of peace enhanced by everything we have experienced in the previous two hours. We see gorgeous birds on our way, including redpoll and stonechat. When I see a stonechat, I am reminded of my youth walking home from school, along winding country lanes with the stonechat as my guide, always sitting on the hedge about one hundred metres ahead, and when I got closer, he would set the next one hundred metres: a pilot friend. And here I am in Scotland, and sure enough, Lesley and I get the same leading spirit. I am sure there is a logical reason, such as steering us away from a nearby nest, however, I prefer the notion of companionship.

As we near Coran, our walk is topped off with a skylark rising and lilting into full song. If you could ask for something to parcel up a memorable day and store it away for keeps, then what better than the generous song of the skylark.

RSPB Nagshead

It is a cold, crisp and sunny January and calling out for a walk. Lesley and I decide to head to the Forest of Dean for a woodland walk, primarily to indulge my hope of seeing a hawfinch as this winter there has been an eruption of these striking birds.

My second and equally strong reason for wanting to visit the forest is my love of woodlands which, along with mountains, are my favourite landscapes. I attribute my appreciation of woodland down to my early years holidaying on my grandparents' farm. Whimple Wood farm was indeed wooded and so much of my time was spent in the goyle, a tree-lined stream running through the farm. Forests and woods are a community, and there is an intimacy that is inclusive and leaves you feeling peaceful and at one with your surroundings.

Arriving at RSPB Nagshead and full of anticipation, we are not disappointed. The weather has prevailed, and there are not too many visitors. The first thing I notice when I step out of the car is the damage caused by the wild boar with all open areas turfed up and churned over. You see reports of the damage on the news, but it is not until you see it first-hand that you can appreciate the bulldozing effect these animals have on an environment. From an ecological perspective, the wild boar is one of the natural inhabitants of our woodlands; however, with such a long absence, their reintroduction and dramatic landscaping comes as bit of a shock.

We decide to take the longer of the two recommended walks. Early in the walk, we stop off at one of the hides that reveals in the distance two fallow does, and as there is little else, we head off on the walk. We both comment on how welcoming the wood feels as the spirit of the place is benign and the mixed deciduous and coniferous trees clearly have an age as some of the tall, graceful broadleaf trees are adorned with old man's beard.

There has been a torrent of rain lately which has resulted in a lot of mud and pooling on the paths. One of the benefits of so much sticky mud is the chance to follow animal tracks and ascertain how fresh they might be. Animal tracks lead from woodland pathways and into the trees on the opposite side of the walkway. It is fascinating to note that the direction of travel from exiting the wood is not always the obvious path taken, instead of which the animals will sometimes travel the footpath for some distance before entering the adjacent wood further along. I can track fallow deer slots and wild boar prints and some birds as they move around the landscape.

Once away from the hub of the reserve and further

into our walk, we have the whole two miles to ourselves, sometimes moving from sheltered wood to quiet glades and open rides. Occasionally, a drift of wind will muscle through, and the trees shiver, with the noisiest being the coniferous as they bustle and jostle high up in the canopy.

Halfway through the walk, we make our way towards a gate with thickset woodland and, beyond, we see more fallow deer, and pride of place amongst them is a wonderful buck with a fine set of antlers. Thankfully, we go unnoticed as they silently move undisturbed through the wood and then, like ghosts, they are gone, as if they were never there in the first place. Great woods and deer belong to each other, a shared and harmonious heritage.

Further into the walk, and we enter a wide, open and sunbathed ride with scrub and an open plantation to our left and a fenced-off section of woodland to our right. As the landscape opens up, this is a fantastic opportunity to see more wildlife and, for the first time, I see a flock of birds land on a tall conifer in the distance but too far for me to get a positive identification. Given the yellow/green plumage illuminated by the light, I think they might be siskin. As I remain still, watching the flock of active birds, I notice movement in the open ride, and there it is: my first wild boar, bristles glistening in the low light. Wild boar have a keen sense of smell, and I have been detected, and as I call to Lesley, the single animal heads for cover, first travelling at an amble and then at speed. I am really chuffed to have seen a boar, and as this was a youngster not yet fully grown, I can only surmise it is a boar driven out to fend for itself.

All around is evidence of the boar and deer, both

in their tracks and through the damage caused to the woodland. Given a natural balance, the churning of the soil and understorey and browsing of the deer should not be viewed as damage but more as the natural cycle of a healthy forest. To effectively sustain a natural environment and prevent overuse by the inhabiting creatures, every animal must have a check and balance, and for the wild boar and fallow deer, this must include a predator that can control numbers, and for me, that really must be the reintroduction of the European lynx. There is no threat to man as the lynx will shy from humans and, given the abundant number of deer and boar, it would have no need to attack livestock. This is a controversial view, but there is so much evidence to prove that a top predator would not just bring benefits to the wood but also the landscape owned and managed by landowners as the hunted will constantly stay on the move for fear of being attacked and therefore not loiter for an extended period in one place, thus reducing extensive damage.

We return to the start, having completed an uninterrupted circular walk, and before heading for the car, we look at the second hide where again little is happening.

This has been a great walk in an atmospheric and characterful wood. The birds have been shy, and sadly no hawfinch, but this does not matter, as the boar and deer have provided so much more. I would definitely recommend RSPB Nagshead to anyone who wants to see a variety of wildlife in a beautiful location.

Mindfulness Moment

It is a still March morning and I find myself once again in the beautiful Highlands, my home from home. Early morning and the tide has turned and is set to rise, so I decide to set out and look for otters. Looking out of the window, conditions could not be better as the water on the Sound of Sleat looks to have melted in the glassblower's kiln, so smooth and silky.

Stepping outside, there is no wind, and as it is Sunday, there are no people, meaning I have the whole place to myself. I have nothing against sharing space, but sometimes it is just lovely to selfishly, and without distraction, have nature to myself.

I decide to head for the slipway where the ferry, during the seasonal months, plies its trade. Today is not one of those days, so no ferry. I deliberately drive slowly to my destination, checking every inlet and small rocky outcrop for the elusive otter, who should by now be feeding on the fresh catch riding the incoming tide.

Reaching the ferry slip road, I park up and get out of the car. The immediate thing that strikes me is the quietude, with every element contributing to the whole. The snow-topped mountains of Skye across the Sound are offset with a backdrop of voluminous clouds. There is no wind, sound or manufactured noise to be heard anywhere. The water very slowly and rhythmically laps against the slipway and surrounding rocks, and beyond, there is a harbour seal spy hopping, inquisitively checking my presence. There are black-headed gulls dancing above the surface of the water like ballet dancers, their feet gliding over the water, and these delicate birds still in their winter gowns occasionally dip to pick up a small breakfast morsel.

Time to pause my unending search for sight and sound and allow the tranquil moment to absorb me. I stand on the pier for an age, emptying my mind of the everyday and mundane and instead become totally captivated by this special moment. I feel totally at one with my surroundings, and a warm peace washes over and through me like the gentle rippling waters at my feet. Time really can stand still as it surely does for me on this occasion.

One of the unending pleasures of nature is being surprised by something remarkable and wonderful, and these moments can only be realised when balanced off against days when an idea does not go as planned. I am pleased I did not see an otter as my desire to record and photograph might have overtaken this wonderful experience. I need to make more of these moments and stop chasing fragments and more frequently immerse myself in the whole.

Triple Buttress

It is an early start on a bleak October morning, and leaving Glenelg, the overcast and grey skies do not look promising for a day's walking in the hills. It is nonetheless an exciting prospect as Lesley and I are with our good friend Cameron who knows the best places to visit and is experienced in the mountains.

We have travelled to the Torridon range for a well-defined climb to the Coire Mhic Fhearchair, which many consider to be the best corrie in Scotland. I am looking forward to visiting somewhere new and gaining elevation with accompanying good views.

Leaving the car parked on the A896 and crossing a small but busy river, we head north with an initial steep section which soon levels out to a steadier walk. The raw weather, lively burns and mist-shrouded, russet-coloured mountains confirm this as one of my favourite times to be in the Scottish mountains.

Aside from the bellowing stags, we initially have the place to ourselves and even throughout the day only see a handful of others. An hour into the walk and we meet three people repairing the path and laying weighty granite stones to protect the surrounding ground from multiple tracks created by stray walkers. It looks like demanding work, but they are cheery and provide a valuable service. A little further along, and despite it still being early morning, we meet a young couple returning from the corrie. These lively youngsters have spent the night wild camping at Coire Mhic Fhearchair and woke to a stag taking a drink at the lochan – now that would be a magical experience, and they were certainly happy recounting the tale.

The rain passes as quicky as it arrives and, in between each burst, the swathe of dark cloud and mountain-hugging mist remains to threaten more; however, that will not prevent us from reaching our goal. Onwards and the path meets the Allt a Choire Dhuibh Mhoir, a river crossed using stepping stones; waterproof boots and a good walking pole help achieve a safe passage.

A little further on and the path takes a fork to the right and includes a steady walk around Sail Mhor. Once we round the corner, we see the waterfall running off Loch Coire Mhic Fhearchair, and the northerly wind is blowing a curtain of water backwards from the uppermost reaches of the energetic fall. Later, and before returning to the car, I refill my water flask from the fall, and it is the loveliest water I have ever drunk, untreated and fresh off the mountain.

Tantalisingly, the corrie and Triple Buttress are not in view, but the impressive waterfall promises much. Before we finally top the path, we meet two ladies of similar age to us who are on their way down, one of whom has just taken a swim in the lochan nestled in the corrie. Their intrepid spirit and bravery of the chilly water give us all hope.

Finally, we arrive at Coire Mhic Fhearchair with the Triple Buttress dominating to the rear. It quickly becomes clear why so many people proclaim this as the most attractive corrie as it has a conspiratorial and beautiful presence. Two steep faces to east and west, great cathedrals of rock that funnel my attention across the lochan to the Triple Buttress. The surface of the lochan is like a millpond and the regimented, large conical piles of scree reflect off the water to the point where it is hard to discern rockfall from reflection. The Triple Buttress are three immense columns leaning as if they are supporting Beinn Eighe, which enigmatically drift in and out of vision as there is a thin and shifting veil of mist across the face and tops of this imposing feature, which might have been a disappointment but in fact is welcome as it adds to the mystery of the place. The muted colours of the land complemented by great slabs of rocks and boulders.

We spend time taking in the atmosphere and imposing presence of the scenery and decide to break out our lunch. Time taken sat with sandwiches and a flask of coffee accompanied by the background soundtrack of the waterfall allows us to appreciate the harmony of the corrie and surrounding mountain. Time spent here with the shifting and changing atmosphere is special and will remain as a treasured memory.

There is a sad footnote to the history of this place as, in the 1950s, a Lancaster bomber was on a flight from the Faroe Islands and crashed close to the summit of

Beinn Eighe, and sadly, all eight on board lost their lives. Wreckage of the plane can still be seen scattered close to the corrie.

Reluctantly, we decide to return the way we came. The changeable and moody weather is still in charge, and rain intermittently makes an appearance, but this will not dampen our spirits. As we descend, rounding Sail Mhor, I look up from focusing on my feet and observe the wider scenery, and there before us the weather is clearing to produce a chequer of light and dark, illumination and shadow. Below and far to the north and east is a broad and expansive rust-coloured glen of weathered bracken and heather, pockmarked by numerous lochs. Standing over this bleak but captivating landscape and set far apart are the majestic mountains of Beinn an Eoin, Beinn a' Chearcaill and others. The land appears vast with no sign of habitation, roads or paths, and as I gaze on with awe and from a safe distance, it is clear I would not want to be lost down there. The sun is now breaking through, and oblique, bright scudding bands of light rush the face of the mountains and slide across the glen. The blanketed dark sky broken up by pockets of blue sky and shafts of bright light, and all of this has resulted in a kaleidoscopic landscape with a clarity seen for miles and as far as Loch Maree. For several minutes we stand in silent reverence and agree how beautiful it is.

It is time to continue our descent as we do not want to lose the light, and the challenging weather will surely return; however, I cannot resist the urge to occasionally look over my shoulder at the beautiful scene we are leaving behind, tugging at me to stay longer and marvel.

It is surprising how quickly the light will recede, and we eventually reach the car as dusk threatens to settle in. This is a nine-mile walk taking us to an elevation of 540 metres, and my feet are telling me it is enough; however, I would not trade an inch for comfort as this is a truly memorable walk. On the way home, we interrupt a stag chasing a hind down a hill – she makes the crossing; he hesitates close to our car; and we stop for an unobstructed view.

Being in the Scottish mountains is a humbling experience as I am always aware of being present on their terms, and I am reminded that my place in the world is quite insignificant. Today is no different; the reward of everything we saw felt like a special privilege, and I know from spending time with Lesley and Cameron, they too feel the same. And speaking to others along the way and listening to their experiences, it is apparent they too have been stirred by this amazing place.

Rumps to Polzeath

Early June and we have been living in Cornwall for five months and been exploring the coast. Today the weather is fair with little wind. The North coast of Cornwall is rugged, untamed and a free spirit and, at this time of year, covered in a great thatch of colour.

Today we have chosen a circular walk taking in the Rumps, Pentire Point and Polzeath, starting at the lead mines near Pentireglaze, above Pengirt Cove. We know it will be a great walk but are not sure what to expect; however, one thing is certain: we are both looking forward to it. Once we get onto the cliff, we are immediately struck by the colour of the sea, with shallower waters resembling the Mediterranean in their colour and luminosity. Contrast this with steep, unyielding dark cliffs, plus drifts of wildflowers, and the visual impact is breathtaking.

The first bird I notice is the kestrel, sometimes referred to as the wind hoverer and for good reason as it faces into the wind and, without effort and making small adjustments, keeps its head stationary and focused on the hunting. This little falcon holds all the finesse that the peregrine, its larger cousin, eschews, whose preference is speed and blunt force. Skirting the lower levels of the cliffs are fulmar, a skilled bird that I am always happy to see. The fulmar masters wind in an entirely different fashion to the kestrel, in that it sweeps up the eddies, drafts and thermals with its stiff and outstretched wings and effortlessly rides them. You can never tire of the fulmar, this great seabird of the northern hemisphere, cousin to the albatross, and these birds are calling as they glide close and flirt with mates perching on the wind-dusted nesting ledges.

Not long into our adventure and we meet a local couple who tell us that this is their favourite walk and they have been doing it for years, telling us we are 'in for a treat'. I cannot wait.

A bird I have seen plenty of this spring is the linnet, and there are numerous on this walk. Early into the walk, a male, with its rose-tinted breast, perches nearby and tinkles its lovely song, giving Lesley her first real opportunity to see this bird displaying its soft summer plumage.

One of the lovely aspects of going for a walk at this time of year is the variety of flowers, and this walk has the bonus of including cliff and coastal flora. On the fields and margins, set back, are the familiar buttercups, foxgloves, oxeye daisies, red campion and bird's-foot trefoil, as well as a wide mix of grasses. Closer to the cliffs and on the clifftops are thrift (sea pink), sea campion, sea carrot and sheep's-bit that hosts numerous insects including the cinnabar moth, a beautiful creature in its striking black and red jacket.

The couple we met earlier told us to watch for the dragon's head and, sure enough, and unmistakably, the nearby headland of the Rumps has all the features of a dragon's head, warts and all.

Never overlook the smallest creatures as they live equally fascinating lives but just not as noticeable. As we continue our walk, I spot a sand digger wasp busy sealing a nest hole with small stones. This is a beautiful wasp with its slim black and red abdomen and just one of several thousand species of wasp found in the UK. The digger wasp, of which there are 110 species in this country, paralyses a caterpillar or moth larvae and drags it to a burrow where it lays an egg on the prey species; it will then seal the nest with sand and small shingle. The egg hatches and the wasp larvae feeds off its paralysed prey. The female wasp will then repeat the entire process. Pretty gruesome but equally fascinating. I stay some time watching and videoing this industrious and beautiful creature as she busily drags small stones to her nest that are half her size and probably as heavy.

Back on the path and I spot a pair of rock dove, the ancestral relative of the town feral pigeon, and these birds set a speedy course across the skyline. This is a bird that relies on speed and dextrous movement to avoid its arch enemy: the peregrine. Over millennia, these two species have been locked in an arms race of pursuit and defence.

We now begin to navigate the Rumps, and right at the beginning of the walk, we meet a couple of birders looking towards a small and inconsequential island called the Mouls. There is nothing insignificant about this wonderful little island as rafting offshore are large numbers of auks which include razorbill, guillemot and puffin. When moving to Cornwall, I never expected to see these wonderful birds off our coast, and they are clearly visible with the naked eye and noticeably clearer using the binoculars. Wherever you get these vulnerable auks, you will always get the avian muscle of the great black-backed gull that are predictably here to predate on the smaller auks. The Mouls is clearly a safe and undisturbed place to make a home. And where sea meets rock, there is an enormous bull grey seal sunning himself.

The Rumps and surrounding cliffs are shored up by steep and precipitous drops to the sea; therefore, a steady foothold and full concentration is required. Whilst we focus on where we place our feet, the gliding fulmar below make a mockery of the threat, and perched on the edge sit busy rock pipits. The views are stunning, and I experience a sense of aliveness, as it feels so good to be in this ancient and rugged landscape surrounded by such amazing wildlife. We must learn to treasure it more. As we leave the Rumps, the walking gets easier, and we see many wheatear and listen to numerous skylarks.

The walk is lovely towards Polzeath, and the views of the town, beach and camel estuary beyond are amazing on this beautiful sunny day. The more distant views to Stepper Point are wonderfully clear. We descend and follow Pentireglaze Haven, a small sandy bay that leads us inland towards the car park. On the final leg, we cross a field of cattle, hawked by swallows and house martins that are feeding on the flies that in turn live off the dung-covered land. The sight of so many hirundines makes me happy as their numbers have dropped substantially over the recent decades, and what is summer without their company.

As we approach the car, a common whitethroat perches on a bramble and gives us a rendition of his song with his little white throat oscillating with each note. To complete the walk, I hear a redstart, not seeing it is not a problem; just knowing it is there is a joy.

Since moving to Cornwall, Lesley and I have enjoyed amazing walks, and this is right up there with the best. It has everything: coast, scenery, wildlife, views and, best of all, unexpected surprises.

Isles of Lewis and Harris

Spending a total of ten days on the Isles of Lewis and Harris in October 2021 and May 2022 has been a wonderful experience offering so much history and natural history.

The evidential presence of people dating back thousands of years can be found everywhere and is especially prevalent on Lewis. The main attraction is the Callanish Standing Stones that predate Stonehenge by approximately five hundred years, and across the landscape are various other neolithic monuments. To visit the main Callanish site at sunset is a special moment as there are very few people present, and the atmosphere is palpable. Permitted to walk among the stones, there is an aura, and the stones exude an energy that is hard to describe. Lesley and I are not alone in feeling those sensations, as there is a hushed silence while the sun navigates the monoliths and settles on the horizon.

It would appear there is a direct relationship between

the main Callanish Standing Stones and Callanish two and three as they are surely aligned. What that relationship is I would not know, but as with the main stones, when at Callanish two and three, there is a wonderful feeling of being connected. It is as if there is a clear link where the past informs the present. Because of the generous energy that is evident at the sites, I cannot help feeling that the people who built these great structures had a superior association and harmonious relationship with Mother Earth and better understood our place in the greater universe than we do today.

When witnessing the human activity on the islands, you do not need to pass too far into the past to find something special, such as the production of Harris tweed. To see the intricacies of the patterns and the history of the crofters producing such beautiful cloth is wonderful, and the production methods have not changed a great deal over the decades. When visiting Gearrannan Blackhouse village, it is humbling to see how the crofters lived, and how the loom had such a larger-than-life presence in an adjoining outhouse. We are given a demonstration of the one-hundred-year-old loom at work, and it is a thing of beauty to watch the various parts perform in harmony, and the noise is bearable as it has a softness that can only be gleaned from old machinery. Without romanticising the past, as with the builders of the Callanish monuments, I gain a sense that the people lived a balanced, albeit hard, life with their surroundings, and for all our modern comforts, we are missing something fundamental and grounding.

Evidence of people leaving the land is everywhere. There are the neolithic monuments, Bronze-Age homes and evidence of the dreadful clearances. There is one other reminder: that of the abandoned crofts with their caved-in roofs and windows. In some, there is evidence of past life, such as wallpaper and kitchen belongings, and in one sense or another, someone has left their home, and it remains, slowly absorbed by the landscape, to remind us that nothing and no one is permanent. I am fascinated by these old, abandoned crofts as they have a sense of place and belonging; they were once receptacles of joy, sadness, tears, laughter, new life and passing, and I wonder, constrained by the boundary of time, who lived there.

Aside from the wonderful history of these islands living on the edge of the Atlantic, there is the landscape and wildlife to marvel at. Whenever I write about being connected and grounded, then this is what I refer to: landscape and nature intrinsically bonded and our place balanced within. It is landscape and nature that are my friends and companions and, on more than one occasion, have been a source of solace, and I clearly feel a strong connection on Lewis and Harris. This is never more prevalent than when the Aurora Borealis makes an appearance. Being on our little balcony and watching the green-gowned 'merry dancers' link arms and lightly step onto the dance floor of the northern sky is not just mesmerising but awe-inspiring and emotional. The cold of an October night holds no authority as the aurora flickers and sways for an inordinate amount of time, and when it is complete, there is no sadness, just a sense of privilege.

The birdlife is abundant, and there are species present that a young birder dreams of seeing. That young birder has grown older and, in retirement, continues to see new and much sought-after species. When starting out as a ten-year-old, I would walk Kingsbridge Estuary, watching and listening to plentiful waders and wildfowl, honing my skills accordingly. In the intervening years, there has been a substantial decline in bird numbers, and the world is a poorer place for the loss, and so it has been joyous to see and hear those considerable numbers still present on these wonderful islands.

To hear the curlew cast forth its plaintiff call has, since those early years, always filled my heart, and this was especially poignant when standing at Callanish two looking to the Callanish Standing Stones witnessing a calling pair glide onto the estuary that borders the two monuments. The melancholic call of the curlew bridges the great expanse of time, and that thread of sound truly fits the landscape, and the sense of belonging is not lost on me: it is meant to be.

Adjacent to our holiday property is Uig Bay, a tidal expanse of beach and adjoining estuary, and this in turn attracts so many species such as greenshank, oystercatcher and common ringed plover. There is one other from my youth that stands out, the redshank, the sentinel of the marsh. This plucky little bird sporting its red leggings, and with its piercing call, is the first to spot danger and warn all and sundry. The redshank was once a common bird and is now, along with the curlew and many others, in serious decline. This beautifully located property is perfect for watching and listening to these estuarine birds and a poignant link to my youth.

A bird that we see regularly is the white-tailed eagle, affectionately referred to as the 'flying barn door', and it is surely a spectacular sight. When a buzzard or raven glides upwards to harry and escort it away, the size difference is remarkable, and yet the buzzard and raven are no small birds. The white-tailed eagle is less wary of humans than its regal neighbour the golden eagle and consequently keeps to a lower altitude; on at least one occasion, it comes within a few metres of our accommodation, its great slab of wing and massive beak casting shadow.

The white-tailed eagle deserves respect and admiration; however, my favourite eagle is the golden eagle, that majestic and mysterious bird of the high mountain. I have been privileged enough to see the golden eagle on many occasions, and every time, I stand in admiration, and the sighting on Harris is no exception. It is a clear, sunny day with excellent visibility, and we are visiting the eagle hide at Glen Meavaig; we wait patiently, also visiting the nearby loch (more of that later) and have our lunch sat outside the hide. The glen is a wide-open track of land surrounded by tall, rugged mountains and, without warning, and from the west, a golden eagle effortlessly rides a high thermal and slowly drifts overhead and disappears eastwards. We are jubilant, and that one distant sighting is worth more to me than multiple sightings of the slightly larger and more accessible white-tailed eagle. The golden eagle is arguably the very symbol of Scottish wildness: mountain, mystery, freedom and proficiency in one majestic animal.

A bird I have always longed to see and hear is the black-throated diver, sometimes referred to as the black-throated loon. This is a striking bird to look at and equally beautiful to hear across a vast and open loch. There are only two hundred pairs nesting in the UK, and they are all found in Scotland, therefore seeing them in this vast and rugged country is not at all easy. On our visit in May, we are extremely fortunate to see three fishing off the coast of Kneep beach which would have been enough to fulfil my ambition, however it gets even better when we see a pair on the small loch adjacent to the eagle hide on Glen Meavaig. This loch gives uninterrupted and close views as they fish the waters. These are striking birds in their black-and-white stripes and prominent black bib with a glistening red eye to finish the look. We do not hear the call but, as with all things birding, be grateful for what you have seen and use anything missed as an excuse to return.

On a dreich, overcast day, we visit Mealasta near Uig bay on Lewis which has a long and varied history, with ruins dating back to the Bronze Age, Highland clearances and, more recently, World War II. We have not long been out of the car when I hear the distinctive and eerie call of the red-throated diver, again sometimes referred to as a loon. It takes a moment to register the call coming from the sea as I am not expecting it, but it is surely a red-throated diver. I love historical sites, especially those unmanaged, but now distracted by this call, I need to see these striking divers! Once I get close to the rocky coastline, they fall silent, and when I move away, they start up again; they are clearly shy and avoid being seen. Eventually, I give up and leave them in peace, and we are in time rewarded by the birds flying overhead, giving their goose-like contact call as they circle towards more open water. As it happens, I do eventually enjoy, and indeed feel touched by, the ruins, and it is palpable that this small and remote spot on Lewis has been truly marked by history, and it is all the more atmospheric when accompanied by the timeless soundtrack of the red-throated diver.

Another bird of significance I will not forget is the merlin. I have seen merlin before on the moors and heathlands in the south, but their numbers are limited. On Lewis, and close to where we are staying, we see them on several occasions; however, the best view is one evening when driving out for dinner and witnessing two male birds in a territorial dispute. They are oblivious to us in the car, and therefore we get fantastic views of their russet-streaked breasts topped off with slate-coloured heads and backs. As the evening setting sky washes soft pink light over the moor and its occupants, the birds posture and move with conviction along the sentried posts lining the road, and we slowly roll alongside to enjoy the spectacle. Eventually, the birds take their argument into the depths of the moor, and we are only five minutes late for dinner – I would miss my dinner for such a sight!

We witness so much wonderful wildlife, including red deer, otter, shags sitting on nests and visiting seabirds such as guillemot and razorbill. I even see my first black guillemot, a small seabird sporting its jet-black coat, white lapels and bright red legs. Another first is the abundant and shy golden plover in breeding plumage, a bird I have seen in the south in its duller winter plumage, but the difference is substantial when seen in its black and gold breeding plumage.

The Isles of Lewis and Harris are stunning and live up to my expectation. I have always wanted to visit the Outer Hebrides as there is something about edge-lands that is mysterious. The people, history, monuments, landscape and wildlife come together to create a land that is more than a set of Atlantic islands set offshore – they are a world separate that demand immersion.

Roughtor Dawn

5.30am and the timepiece has woken us from a deep September slumber. The season is changing, and darkness keeps a tighter grip on the clock. A quick wash and a brush of the teeth and Lesley and I head out for a small local adventure.

I have, for months, wanted to photograph the sunrise over Bodmin Moor from the high vantage of Roughtor. However, to achieve this ambition, it requires an early start and a walk across the moor in the dark.

The drive to Roughtor car park is approximately five minutes and, on the journey, the car headlights pick out two tawny owls, denizens of the night and most definitely my favourite of our native owls. Once we have parked, we switch on our torches and set out for Roughtor. The sun is due to rise at 7am and already there is the hint of dawn in the eastern sky. To the west languishes Jupiter, the bright and influential guardian of our solar system, and riding high in the sky is our nearest neighbour, the unmistakable red planet of Mars. Sitting just below Mars is Orion, that great hunter of the sky, wearing his bejewelled and shimmering belt. The cloudless sky is beautiful with its soft pink to the east, inky darkness to the west and, throughout, studded with a multitude of unblemished stars and planets.

An intriguing and time-bound aspect of the moors in the south-west are the neolithic and Bronze-Age ruins that are apparent and accessible to walk amongst. I find these old sites fascinating as they are thousands of years old and not just a reminder of past lives but also the grounding impermanence of my own existence. During the day, I am acutely aware of the Bronze-Age dwellings and enclosures, and I am surprised by my heightened sensitivity as, in the dark, I walk in their presence, a sense of community.

Climbing higher and trying to stay on track, my torch picks out the shadowy forms of the moorland sheep that quietly mind their own business, grazing as they do so. Looking up to my destination, the sky is marginally brightening, and the imposing form of Roughtor stands proud. From previous visits, I know the exact location to capture the best image of sunrise, and I head for that landmark, allowing myself enough time to set up my equipment. It is at this point that Lesley and I take different routes to the top; as I follow the narrow sheep tracks, Lesley takes the circuitous and more clearly defined path. At the top, the sky to the east is gaining colour and deepening in intensity and Brown Willy, slightly to the south-east, looks fearlessly majestic. Near and far, the low-lying land is enveloped by a deep mist, and the gentle hills roll over the swell of glazed fog like boats setting out from port.

I am joined by two photographers from North Devon who arrive twenty minutes after me but set out one hour before. They have Dartmoor and Exmoor at their disposal but want to try somewhere different. Whilst we share best ideas on where and how to photograph the landscape, Lesley quietly sets off to explore for herself.

The sunrise is beautiful, and the colours are amazing. A few minutes into the sun's early climb, the casting light picks out focal points on the landscape, and deep pools of shadow offset the warming highlights; down in the valleys, the brightening seas of mist drift across the land.

The granite columns, slabs and boulders dominate the moor, and as I face into the sun, the deep scars and smothered fissures are shadowed to one side and highlighted on the other. It is from one of these deep recesses that a merlin without warning makes a quick dash and skirts the rocky outcrop, flying away from me, in pursuit of breakfast.

Today is the state funeral of the Queen, and I can think of no better place or time of day in which to reflect on her passing and the great service she gave to the nation.

The time has come to depart; we have had a wonderful morning, and it is still early. Heading down the moor, we find ourselves in a boggy patch with a trickle of water running through it, and we are thankful we did not stumble into this when dark.

This walk has it all: celestial bodies, presence of ancient abodes, moorland sunrise and birds of prey. Being in the country before dawn and experiencing all that we have today cannot be replicated at other times as there is a freshness that arrives with the new day, and the stillness is only disturbed by natural elements.

Two Walks, Two Falcons

Mid-January 2023 and finally we are getting some clear days free of torrential rain and battering wind. It's a strange phenomenon that, when in Scotland, I will walk in all weather, but when at home, I have become a fair-weather walker. It was not always the case, and when out in poor weather, I often feel more alive and connected to my surroundings. Perhaps it is an age thing!

The first of the two memorable walks is a circular one leaving from home. We plan to cut across the fields to Treclago Farm and then take the road and lane to the Watergate Road, onto Pencarrow and home. A reasonable stretch and one of my favourite walks.

The steep field leading to Treclago Farm is inhabited by two large and inquisitive horses grazing on the far side of the field. Their curiosity gets the better of them, and they trot across. Despite my rural upbringing and farming ancestry, I am not a huge fan of large livestock, and these two are now up close and in my space and, through excitement, one of them kicks out its large back feet; the gate can't come soon enough!

The traffic-free track leading to the farm is broad, a mix of grass and mud, and the sun illuminates the deep hedge with its early signs of the spring to come. Turning left at the farm, we take to the narrow lane that has no through route and leads to Aldermoor Farm. Along this old lane lined with aged Cornish hedge and overhanging oak, there is evidence of long past habitation and fields that are marked with old trees and scrub. It is one of the most peaceful lanes I know of. Once at Aldermoor Farm, we head up the old drover's lane that has broad verges where native plants are left to propagate in the dappled light, and at this time of the year, chief amongst them is the yellow gorse, blooming shades of sunshine and producing a sweet scent, a timely provider of nectar for any insect that might awake early.

At the top of the drove is a crisp view of Roughtor, dominating the horizon. In the field adjacent to the road are some beef cattle feeding on silage with a haze of steam rising from their hides and the same inescapable moisture billowing from their flared nostrils.

It's not long before we reach the ford where the adventurous take their vehicles through the clear, rushing water. For the more cautious, there is a small bridge with battered rust and green painted railings lining the way. It is here that we stop to view a flock of two hundred lapwing, along with a few golden plover, that have taken up residence in a sodden field. Half the birds are grounded and settled whilst the remainder appear troubled, sweeping and arcing across the sky, seemingly unable to make the ground.

Around they go multiple times, calling *peewit* throughout the balletic movement. In flight, the lapwing are beautiful birds, their broad wings creating a bounce with each stroke. The sun glistens on the grounded birds, and the purples, blues and greens shimmer from the lapwing plumage. In comparison, the golden plover are without their burnished gold and black overcoats until the spring demands they show off.

Dragging ourselves away from this rare and beautiful spectacle, we head up the road towards Pencarrow. We haven't long continued our journey when there is a mass disturbance, and all the lapwing and golden plover are in the sky, creating a mass confusion of flight and noise. Something has spooked them, which very soon becomes apparent as a merlin in pursuit of a small bird travels at speed along the hedge adjacent to where we are standing. As with all things merlin, it happens very quickly, and as this speedster has come from the direction of the lapwing, it is safe to assume this smallest falcon is the cause of so much disturbance. Good luck little passerine being chased down.

Onwards and it is not long before we see an old friend. For three years, we have walked this route, and on a specific telegraph post sits a buzzard, and it will sit tight until we get within three hundred metres. It may not be the same bird, but as they have an average lifespan of eight years, it most probably is the same buzzard. As it nonchalantly drops off its standing post and wheels across the open field, we look forward to seeing it again when next on this road.

Eventually, we arrive at the River Camel and our home leg. The river is in full spate and in an unforgiving mood. There is the roiling mass of churned water with the occasional deep pool and sheltered water set behind rocks or fallen log. It is on one of these hidden sanctuaries that we see the dipper bobbing on a rock. I never tire of seeing our only aquatic songbird with its chocolate-coloured plumage and white bib.

The weather is still inviting, so two days later, we set out on a coastal walk. We decide on a circular walk from Tintagel church to Trebarwith Strand. The low sunlight is fiercely bright, reflecting and settling on the damp pathways like liquid mercury. The sea still retains some energy from past storms and reels coastwards in great tubular rolls; once there, it crashes against the rock-defended cliff face. On impact, there are great spumes of spray sent skywards and, before descending, it leaves behind a fine lingering mist that glistens on the penetrating light. On the leeward side of the collision, the cliff retains some clarity, resulting in a concertina of illuminated and shrouded rock face. This early sight is mesmerising and stays with us throughout the walk.

Not for the first time on this walk, I see a peregrine. Judging from his size, this is a tiercel, and I have a great view as I look down on him whilst he nonchalantly powers his way up the coast towards Tintagel. Even in this apparently relaxed state, it exudes power. I have seen peregrines in city centres, Somerset Levels and numerous occasions on the coast, and it is here that I most enjoy seeing them. Coastal cliffs are the peregrines' natural and original habitat where they hunt, amongst other prey, the rock dove. It is no surprise that the peregrine has moved into the urban environment where it can hunt the abundant feral pigeon, relative of the rock dove.

Further into the walk and we approach Lanterdan Quarry. I have visited this quarry on a few occasions, and I am always struck by the impact of the quarrymen on the cliff face. There has been 650 years of quarrying on this spot, resulting in a deep gouge taken out of the cliff, and standing sentinel is a large column of rock that was of no commercial value to the workers, so they toiled around it. There is a deep sense of presence here, the hardships the quarrymen must have undergone in all weathers to extract the slate.

Continuing along the path, and within sight of Trebarwith Strand, we take a narrow drove inland and make our way back to the car. The light is still brilliant and the walking easy as we make the final stretch towards the church. In the field to our left are a small flock of twenty starlings, heads down and busy foraging, and close by is a solitary crow that is lazily strolling around looking for easy pickings. The hedge to my right is not particularly high, and I can see the sheep grazing in the field.

Without warning, to my right, I catch sight of a peregrine flying in fast and low, lifting itself over the hedge to surprise the starlings in the adjacent field. The crow coughs a warning, and the starlings bounce upwards as one. This confusing mass is off-putting, and the muscle-bound falcon misses out on lunch. Having launched and completed its surprise, the peregrine eases its way towards the cliffs, slowly enough for me to get a good view with my binoculars. This again is a tiercel, perhaps the same bird I saw earlier in the walk? I have watched peregrines launch attacks from high on an arrow-like stoop, and I have also seen them harry flocks of waterfowl on the

Somerset Levels, but this is the first time I have seen them use ambush tactics which are more synonymous with the sparrowhawk or goshawk.

Over the last forty-eight hours, we have seen Britain's biggest and smallest falcon, both in hunting mode, and they are spectacular. The gutsy little merlin that punches above its weight, relentlessly chasing down prey, and the bigger peregrine, confident, powerful and resourceful in its technique. As they hunt in their own niche and without competition of resources, it is possible to see both within miles of each other. When they are perched, it is possible to observe similarities but, equally, there are striking differences in plumage and, in the case of the merlin, a restlessness, leaving little time to admire their beauty. I have been fortunate to see both birds on numerous occasions, and they still create a stir of excitement.

Acknowledgements

To Jill, my mother, for hanging out the bird feeders early in my life and introducing me to the fascinating world of birds. To my stepfather David for all the farming knowledge. To Roy, my father, for his love of photography that I absorbed from an early age. To my grandparents, Gag and Bappy, for allowing me the freedom to explore their farm, my little piece of heaven.

A big thank you to Lesley for sharing my love of the countryside and wildlife and for patiently, in all weathers, waiting for me to set up camera equipment, sketch and note the scenery and wildlife around me.

Thank you to Sam, Daniel, Lauren and Fiona who sometimes share my passions and occasionally seek me out for advice on wildlife identification.

To all my friends who share my joy of the natural world and sometimes join me on my walks. A special thank you to Cameron for introducing me to some Scottish gems that

I would never have found for myself.

To all landscape and wildlife writers, artists and photographers: keep creating and shine a light on our beautiful world and hopefully bring others on your journey.